TALES FROM THE
ANT WORLD

ALSO BY EDWARD O. WILSON

Genesis: The Deep Origin of Societies (2019)

The Origins of Creativity (2017)

Half-Earth: Our Planet's Fight for Life (2016)

A Window on Eternity:
A Biologist's Walk Through Gorongosa National Park (2014)

The Meaning of Human Existence (2014)

Letters to a Young Scientist (2013)

The Social Conquest of Earth (2012)

Why We Are Here: Mobile and the Spirit of a Southern City,
with Alex Harris (2012)

The Leafcutter Ants: Civilization by Instinct,
with Bert Hölldobler (2011)

Kingdom of Ants: José Celestino Mutis and the Dawn of Natural History
in the New World, with José M. Gómez Durán (2010)

Anthill: A Novel (2010)

The Superorganism: The Beauty, Elegance, and Strangeness of
Insect Societies, with Bert Hölldobler (2009)

The Creation: An Appeal to Save Life on Earth (2006)

Nature Revealed: Selected Writings, 1949–2006 (2006)

From So Simple a Beginning:
The Four Great Books of Darwin, edited with introductions (2005)

Pheidole in the New World: A Hyperdiverse Ant Genus (2003)

The Future of Life (2002)

Biological Diversity: The Oldest Human Heritage (1999)

Consilience: The Unity of Knowledge (1998)

In Search of Nature (1996)

Journey to the Ants: A Story of Scientific Exploration,
with Bert Hölldobler (1994)

Naturalist (1994; new edition, 2006)

The Diversity of Life (1992)

The Ants, with Bert Hölldobler
(1990; Pulitzer Prize, General Nonfiction, 1991)

Success and Dominance in Ecosystems:
The Case of the Social Insects (1990)

Biophilia (1984)

Promethean Fire: Reflections on the Origin of the Mind,
with Charles J. Lumsden (1983)

Genes, Mind, and Culture, with Charles J. Lumsden (1981)

On Human Nature
(1978; Pulitzer Prize, General Nonfiction, 1979)

Caste and Ecology in the Social Insects,
with George F. Oster (1978)

Sociobiology: The New Synthesis
(1975; new edition, 2000)

A Primer of Population Biology,
with William H. Bossert (1971)

The Insect Societies (1971)

The Theory of Island Biogeography,
with Robert H. MacArthur (1967; new edition 2001)

TALES FROM THE
ANT WORLD

EDWARD O. WILSON

LIVERIGHT PUBLISHING CORPORATION
A DIVISION OF W. W. NORTON & COMPANY
INDEPENDENT PUBLISHERS SINCE 1923

Frontispiece: An African warrior ant, or matabele (*Megaponera analis*), specialized for raids on mound-building termites. Its common name is that of the Zimbabwe warriors Long Shield. *(Original painting by Timo Wuerz.)*

For information about permission to reproduce selections from this book, write to Permissions, Liveright Publishing Corporation, a division of W. W. Norton & Company, Inc., 500 Fifth Avenue, New York, NY 10110

For information about special discounts for bulk purchases, please contact W. W. Norton Special Sales at specialsales@wwnorton.com or 800-233-4830

Manufacturing by LSC Communications, Harrisonburg
Book design by Patrice Sheridan
Production manager: Julia Druskin

ISBN 978-1-324-09109-7 pbk.

Liveright Publishing Corporation
500 Fifth Avenue, New York, N.Y. 10110
www.wwnorton.com

W. W. Norton & Company Ltd.
15 Carlisle Street, London W1D 3BS

1 2 3 4 5 6 7 8 9 0

CONTENTS

INTRODUCTION:
ANTS RULE

EVERYONE LIVING OUTSIDE the polar ice sheets who has gazed around their own feet has seen ants, and inevitably, they have heard tales about these social creatures, especially concerning their relationship with humanity. Ants, it is said, are among the little creatures that run the world, perhaps for our benefit, or perhaps not. Ants form societies that rival roughly in form and variety those of human beings. Then, there is their awesome abundance. If *Homo sapiens* had not arisen as an accidental primate species on the grasslands of Africa, and spread worldwide, visitors from other star systems, when they come (and mark my word, they will eventually come), should be inclined to call Earth "planet of the ants."

I've written *Tales from the Ant World* from the experience of a lifetime, some eight decades, of studying fabulous insects. I began in grammar schools in Washington, DC, and Alabama, and continued, with the same emotions, to university research professor and curator of entomology at Harvard University. In these *Tales,* I convey some of the importance of what I have learned from my studies and those of others. Incidentally, I and my colleagues are called myrmecologists in the scientific academy. And even though I have now written over thirty books, most of them scholarly, I have not until this book told the amazing stories of myrmecology as a physical and intellectual adventure—if you will, an adventure story.

I am especially hopeful that this account will reach students—even ten years old is not too young—interested in the prospect of a scientific career. The subject at hand is wide open. The existing natural history and biology of ants covers only a tiny fraction of the more than 15,000 ant species discovered to date, given a name, and studied carefully. And beyond the ants, more than a million species of insects, spiders, and other arthropods await full attention. The more that this part of the biosphere is studied by future experts, the better off will be the world, ourselves included.

Meanwhile, the most frequent question I am casually asked about ants is, "What do I do about the ones in my kitchen?" My answer is, Watch where you step, be careful of little lives, consider becoming an amateur myrmecologist, and contribute to their scientific study. Further, why should these wondrous little insects *not* visit your kitchen? They carry no disease, and may help eliminate other insects that *do* carry disease. You are a million times larger than each one. You could hold an entire colony in your cupped hands. You inspire fear in them; they should not in you.

I recommend that you make use of your kitchen ants by feeding them and reflecting upon what you see, rather like an informal tour of a very foreign country. Place a few pieces of food the size of a thumbnail on the floor or sink. House ants are especially fond of honey, sugar water, chopped nuts, and canned tuna. A scout in close vicinity will soon find one of the baits and, to the degree the colony is hungry, run excitedly back to the nest. There will follow social behavior so alien to human experience that it might as well be on some other planet.

TALES FROM THE
ANT WORLD

1

OF ANTS AND MEN:
MORALITY AND TRIUMPH

I'LL BEGIN THIS myrmecological tour with a word of caution. There is nothing I can even imagine in the lives of ants that we can or should emulate for our own moral betterment.

First, and most importantly, all ants active in the social life of colonies are females. I am an ardent feminist in all things human, but in ants one has to consider that during their 150 million years of existence, gender liberalism has run amok. Females are in total control. All ants you see at work, all that explore the environment, all that go to war (which is total and myrmicidal) are female. Adult male ants are pitiful creatures by comparison. They have wings and can fly, huge eyes and genitalia, and small brains. They do no work for their mother and sisters, and have only one function in life: to inseminate virgin queens from other colonies during nuptial flights.

To put the situation as simply as possible, males are

little more than flying sperm missiles. Once launched, they are not allowed to reenter their home nest, though if successful they may become fathers of new colonies comprising, in some species, many millions of daughters and sons. Whether successful or not at reproduction, they are destined thereafter to die within hours or at most a few days from rain, heat, or the jaws of predators. They cannot just stay at home. They do no work there and are otherwise a burden on the colony. After the nuptial flights, should they linger they are driven out by their sisters.

Second on the list of formicid moral precepts, after absolute female rule and much more horrific, many kinds of ants eat their dead—*and* their injured. If you are an elderly or disabled worker, you are programmed to leave the nest and not burden the society any further. If you die while in the nest, you will be left where you drop, even on your back with all six legs sticking up in the air, until your body emits the odors of decomposition, which are, in particular, oleic acid and its oleates. When you smell dead, your body will be carried to the colony refuse pile and dumped. Or, if just mangled and dying, you will be eaten by your sisters.

In a third morally dubious propensity, ants are the most warlike of all animals, with colony pitted most violently against colony of the same species. Extermination is the goal for most, and as a rule larger colonies defeat smaller ones. Their clashes dwarf Waterloo and Gettysburg. I have seen battlefields strewn with dead warriors, a high percentage of which, as it turns out, are aging females. As adult workers grow older, they take on increasingly dangerous activities on behalf of the colony. At first, most serve as attendants of the queen mother and her brood, from eggs to larvae and through pupae to newly emerged adults. Next they engage more in nest repair and miscellaneous other internal tasks. Finally, they become prone to service outside the nest, from sentinel to forager, to guard, to warrior. In a nutshell and put more plainly, where humans send their young adults into battle, ants send their old ladies.

For ants, service to the colony is everything. As individual workers approach natural death, it benefits the colony more for the old to spend their last days in dangerous occupations. The Darwinian logic is clear: for the colony, the aged have little to offer and are dispensable.

Evolution at the level of organized groups has paid off richly for the more than 15,000 species of the world's ant fauna. Ants are the dominant land carnivores in the weight range of one to one hundred milligrams. Termites, sometimes erroneously called "white ants," are the dominant consumers of dead wood. Together, ants and termites are the "little things that run the world," at least among the animals of the terrestrial world. In the Brazilian rain forests, for example, they make up an astonishing three-quarters of the insect biomass and more than one-quarter of the entire animal biomass.

Ants have been prominent on Earth over one hundred times longer than have humans. They are estimated (by molecular methods) to have originated about 150 million years ago. They then diversified into myriad anatomical forms 100 million years before the present, by the end of the Age of Reptiles. A second such radiation occurred during the early Age of Mammals. The modern species *Homo sapiens*, in contrast, emerged in Africa only one million years ago, a fraction of that time.

Had extraterrestrials visited Earth at any random time during the past hundred million years, they would have discovered abundant life clothing the land. They would have found the fauna and flora dominated by ants and thereby, in part, healthy and intact. The extraterrestrials would have become myrmecologists. They would find ants, as well as termites and other highly social creatures, somewhat bizarre, yet for that reason a key force in sustaining stability in almost all the terrestrial ecosystems of the planet.

The extraterrestrials might then report to their home planet, concerning Earth, "All is in order. For a while."

2

THE MAKING OF A NATURALIST

NATURE IS THE metaphorical goddess of all existence that lies beyond human control. Humanity is blessed to the extent we love her, and her products, from the sweet descent of her sunsets to the tantrums of her thunderstorms, and from the empty vast space beyond her biosphere to the seething diversity within it, of which we ourselves are a recent chance addition.

The love of Nature is a form of religion, and naturalists serve as its clergy. The goddess, we believe, will lead us from the darkness into light. For those who follow, she has made the ultimate promise of all religions: grant Nature eternity on this planet, and we as a species will gain eternity ourselves.

My own life is the result of an early blend of the two faiths, the first traditionally pious and the second scientific. I consider myself fortunate, in that during the years I spent laboring through public schools, I spent most of my time preparing for a career in natural history. I wanted within all my dreams to be a professional naturalist. I never gave any other option a second thought. As a result I paid little attention to classwork, sports, and social activity.

This slighting of normalcy was due in part to the odd circumstance that I was the only child of four parents, attended sixteen schools across eleven grades in as many towns and cities, and was subject while growing up to the confusion this multiplicity inflicted. My father, Edward, Sr., and my mother, Inez, divorced when I was eight years old. During this drama, unusual for the 1930s, I was parked for a term in the famously strict Gulf Coast Military Academy, since closed. I was next placed in the paid care of a grandmotherly lady, Belle Robb ("Mother Robb"), who was wonderfully kind to me. She was also an excellent cook, notably for her uniquely tasty fried grits cakes. Mother Robb was the best of all possible child guardians, from a young boy's point of view: she let me do mostly what I pleased. There was one exception: that I take an oath to God that I would never drink alcohol, smoke, or gamble. And, above all, I must swear to love Jesus with all my heart and soul. Our Savior, she assured me, would return personally to visit me once in a while. When I grew impatient waiting for the physical Jesus, Mother Robb conceded that He might appear as no more than a flash of light in some place like an upper corner of my room.

In time, the failure of a personal Second Coming didn't matter. I had developed other interests. With Mother Robb's encouragement, I made a collection of every kind of insect I could find around the neighboring houses, in the vacant lots, and along the streets from the Robb home at 1524 East Lee Street, Pensacola, Florida, thence to the local grammar school I attended. It was an exciting adventure for a child my age, one I still pursue on a larger scale, and a forerunner of an important data-gathering procedure in modern ecology, the All Taxa Biodiversity Inventory (ATBI). I watered tropical plants Mother Robb kept on the porch and everywhere throughout the house. I fed my pet baby alligator, and began digging a hole in the backyard I hoped would take me all the way to China.

But beyond being a typical boy, my most influential activity was made possible by a Christmas gift from my mother: a child's compound microscope. With this instrument I spent hours watching rotifers, paramecia, and other microscopic organisms abundant in drops of pond water. This adventure was to work a powerful influence on me for the rest of my life. I have never changed: I experience a similar thrill whenever I visit unfamiliar habitats in different parts of the world in search of new kinds of plants and animals.

In 1939, when I was ten years old, I left Mother Robb and Florida and rejoined my father, by then a government employee with a new wife, my stepmother Pearl, to live in an apartment on Fairmont Street in Washington, DC.

It was one of the happiest chance events of my life. I found myself now living only five city blocks from the National Zoological Park, informally called the National Zoo. Just beyond this wonderland, filled with large animals from around the world, lay the woodland and pastures of Rock Creek Park.

Inspired by field guides and dazzling photographs in library issues of *National Geographic*, free entrance to the National Zoo, and the wilds of Washington, DC, to explore, I became a fanatic about butterflies. By taking time out from schoolwork, I was able to build a large collection. My main instruments were pins and specimen boxes, and an insect net made for me by Pearl. (In later years and new adventures, I found it easy to make a net quickly, anywhere. Take a sawed-off broomstick for a handle, bend a coat hanger to form a circle and attach it to the handle, and sew cheesecloth into a bag hung from the coat hanger rim.) I became adept at locating and netting almost all the many species that fly within and around our nation's capital. And to this day, I remember them all in vivid detail. There were fritillaries abounding in front yard gardens, red admirals chasing one another in territorial battles back and forth around parked cars, tiger swallowtails speeding overhead, what may have been a giant swallowtail (but flew into a tree canopy and eluded me), countless sulfurs, blues, hairstreaks, cabbage whites, and (a triumph!) one specimen of the native white species. I sought but had not a single glance of the mostly winter-dwelling mourning cloak.

Give me today a butterfly net and a spring and summer in the nation's capital (and a letter to show the DC police) and I believe I could happily repeat my adventure.

My fascination with the natural world began to spread with time and more subjects and places to explore. I was aided by my best friend, Ellis MacLeod, who twenty years later would become a professor of entomology at the University of Illinois (at the same time I received a simi-lar appointment at Harvard). As boys together we took an interest in ants. The source of our inspiration was a *National Geographic* article entitled "Ants: Savage and Civilized," by William Mann, the director of the National Zoo at the very time I was making frequent visits to see the large animals and net butterflies in the Zoo's gardens. To continue the coincidences, Mann had earlier earned his PhD under William Morton Wheeler, professor at Harvard, and my predecessor as curator of entomology and builder of the ant collection (with Mann's help) of Harvard's Museum of Comparative Zoology.

In "Ants: Savage and Civilized," Mann gave an account of species found in mostly tropical countries. It soon dawned on us ten-year-olds, Ellis and me, that the only ant he depicted that we could hope to find in Washington, DC, was the "Labor Day ant" (scientific name *Lasius neoniger*), whose small crater nests abound in yards, gardens, and golf courses throughout almost the whole of the eastern United States, and whose common name derives from the swarms of males and virgin queens emerging to mate following a heavy rain within a week or so around Labor Day.

This nascent interest was then cut short. After two years' residence in Washington, DC, our little family migrated back to Mobile, Alabama. It was home, where almost all of my father's ancestors had lived since the 1820s. My paternal grandmother, Mary Wilson, had died, leaving my father and his brother, Herbert, the large house built by my great-grandfather.

Fortunate once again, I found myself a short distance from unusually rich natural environments, this time overgrown vacant lots along with the remnant marshes and woodlots lining the dock area of Mobile Bay. With a new, balloon-tired Schwinn bicycle, I could easily reach a rich mix of wild and semi-wild habitats as far as the Dog River and Fowl River crossings, on the road to Cedar Point, and at the end an unpaved road to the dock for transport to Dauphin Island. My experiences with butterflies and ants deepened, and my interests broadened to include many other kinds of insects. There was also my newest love, the snakes and other reptiles that abound in variety along the coast of the Gulf of Mexico.

My general drift toward becoming a naturalist was deepened and set firmly by yet another emigration, this time to the little Alabama town of Brewton, located close to the Florida Panhandle border just north of Pensacola. Bucolically pleasant in both its people and dwellings, with a population stable at about 5,900, surrounded by "swamps"—floodplain forests dissected by freshwater streams—it is part of the central Gulf Coast region recognized today as having possibly the richest terrestrial animal diversity in North America. Therein dwell thirty-two species of snakes; fourteen species of turtles (a fauna rivaled only by the Mekong delta and parts of the Amazon watershed); dramatic arrays of freshwater fishes, crayfishes, and mollusks; plus everywhere seemingly boundless ants, butterflies, and other insects.

Decades later, I used Brewton as my model for Clayville, an imagined Southern town, in my novel *Anthill*. (This work, to my pleasant surprise, was given the 2010 Heartland Prize, for best novel on American life.) Brewton, for its part, rewarded my admiration by naming a nature park after me. The reserve is relatively large, stretching from within the town limits in one direction toward Burnt Corn Creek, where during the War of 1812 Red Stick Creek warriors defeated a contingent of Alabama militia, and in another direction toward Murder Creek,

where bandits robbed and killed a group of early Brewto-
nians on their way to Pensacola to purchase bullets.

I gained some credence among my teenage peers by
being the first Boy Scout in the surrounding area to earn
the rank of Eagle. Also, for sitting on the bench as the
third-string defensive end of the football team. (I was
called to play only once, in the final minute of the winning final game, with the proudly remembered words,
"Wilson, take left end.") And finally, for hand-capturing
venomous cottonmouth snakes and exhibiting them to
my fascinated peers. (The Wilson method, which I recommend here only for the experienced adult, is the following: let the snake start to move away from you, pin
it with a sawed-off broomstick as safely as possible, close
to the head, roll the stick forward to pin the entire head
firmly, pick up the snake just behind the head, raise the
entire snake with your loose hand and drop it into an
already opened sack.) Amid boys my age with nicknames
like A.C., Chip, Buzz, and Rusty, I was rewarded with
one of my own, Snake. This was all a "Southern thing,"
as they say. The same epithet was later earned by a professional football running back for his skill at weaving
through lines of the defending team.

3

THE RIGHT SPECIES

IN THE SUMMER of 1945, shortly after my sixteenth birthday, my father moved our little family from Mobile, Alabama's coastal city, 337 miles north to Decatur, where the Tennessee River crosses the north-central tier of Alabama counties. There, within and all around the little port city, a new natural world opened up to me. It would have an especially profound effect on my life and scientific career.

Edward, Sr., was a professional traveler, a financial auditor for the Rural Electrification Agency, which supplied electrical power to towns and farms across the rural South. In order to stay close to both work and home, he chose to move our residence every several years.

One result of our peregrine existence was that during my public education I attended fifteen schools in sixteen towns and cities in three states plus the District of Columbia.

A schedule of this withering complexity can be rough

Ed Wilson at 14, in his butterfly, dolichopodid fly, and snake period, before switching to ants as the favored group.

on an adolescent. I adapted by turning even farther toward Nature. Where I found it difficult to make new friends, and thereby gain entry into the usual teenage friendships, cliques, and athletic teams, I turned instead to natural habitats to find a reliably familiar environment.

Hence I spent summers mostly on my bicycle, searching for scraps of wild environment that had survived within and around Decatur and beyond, into the copses and old-growth fields. The wildest places were across the Tennessee River, many abandoned during the Second

World War, which still raged abroad. I seldom saw another person, and then only at a distance.

On the Decatur side I discovered a natural cave, and despite a mild case of claustrophobia explored it in search of blind, white crayfish and other troglobites, as such underground specialists are scientifically designated. Fortunately I never lost my way in the cave. No one would have known where to look for me until my bicycle was noticed outside the cave entrance. But I was never lost, at least for no more than a few hours, and never have been while exploring other wild environments.

Today, writing seventy-five years after my arrival in Decatur, I perceive it as fortunate that I took little interest in making friends or in any way being popular or even winning acceptance in the social life of the Senior High School. My paramount and almost exclusive ambition was to become an expert in some aspect of natural history, and to learn the science supporting it. I was preparing for a university education. My room in Decatur was filled not only with standard high school references and textbooks but with field guides to the plants and animals of North America.

Literally millions of species existed around the world in natural environments, and still do thus far into the twenty-first century. The study of any one species, I understood, could provide the beginning of a scientific career. The challenge to me in my teens, as it is for most acolytes of my kind, was not to discover the best way to study any particular chosen species or cluster of species. Rather, it was to choose the *right* species. I understood that if I chose wisely, my career as a scientist could begin as early as my first year of college.

I needed courage and ambition to get started. The very idea of higher education was daunting, especially since I would be the first in the history of my family on either my father's or mother's side to attend college. Furthermore, money was scarce. My father was ailing, a result of his service in World War I; it was in the army that he learned his trade as an accountant. Given that he had received no more than a seventh-grade education otherwise, I came to admire him for the steep hill he climbed to achieve his profession in life. His example gave me added strength in my determination to do whatever it took to become a professional scientist and naturalist. I thought I needed to go to a university, although in later years I became persuaded that a good liberal arts college probably would have done as well.

I was especially attracted to the scholarship program at Vanderbilt University, advertised during my senior year of high school in Decatur, as it seemed conducive to my ambition. I even took an examination the university had supplied. I tried hard, but was not awarded a fellowship or even an unaided admission. Many years later I mentioned my failure, without malice, when I gave the inaugural address at the opening of Vanderbilt's new Science Center.

Like millions of other youngsters in America, I came to realize that my only resource for attending college would be the meager amount my parents could provide, plus whatever I could earn and save on my own. So during my senior year of high school I took whatever job I could find and saved my earnings. I moved from one to the other to make the most: a paper route, next a door-to-door magazine salesman, then a soda jerk in a five-and-dime store, a stock clerk in a department store, and finally, an office boy in the local steel mill. I was desperate to succeed, and I did well. When at the end of the summer of 1946 I prepared to leave the steel mill, the manager of the division I served said, "Ed, you don't have to go to college. Stay here at the mill. You have a high school diploma! You'll go far."

I decided not to discover how far. In any case, the Alabama legislature saved me. At least, its members took action that accidentally admitted me to the University of Alabama. They evidently foresaw that campuses, with the end of World War II and passage of the GI Bill guaranteeing college-level educational support for those who had served, would be swamped with returning veterans. They passed a law directing that anyone who was a resident of the state and a high school graduate would be admitted to the University of Alabama. I was not a veteran, but I passed the other two basic requirements, and upon application was admitted to the state's premier university. It should be no surprise to anyone that to this day I have remained one of the University of Alabama's most loyal alumni, in every way possible.

Now came the big choice. With the University of Alabama in my future, what should be the insects—whether a group of related species or just a single species—on which I could become the world authority, and thereby launch my career as a scientific naturalist? I made field trips into the fauna and flora all around Decatur. I fished in the nearby Tennessee River, memorized the names of every local fish species, from perch to alligator gar, hunted for snakes in the local woodlands, searched in my nearby cave for blind white beetles and iron-like mandibulate carnivorous crickets. And I undertook on my own a college-level study of insects in general.

An early candidate was a bizarre set of species that forms small ecosystems likely to date back in prehistory for hundreds of millions of years. Sponges are everywhere in the world, primarily as dwellers of shallow seas, but here and there can be found species that live in freshwater streams and lakes. In one pristine brook on the edge of a farm I found colonies growing in thick beds on the bottom. They formed a unique ecosystem, with evidence of damage from caterpillar-like insect larvae with the common name spongillaflies, also of ancient origin. Sponges and their parasite flies: amazing! But I decided not to choose sponges and spongillaflies as my favored group for study in college and beyond. They are rare, and obviously hard to find just anywhere.

Not so dolichopodid flies, the "long-legged ones" as their Latinized name translates. I found them everywhere

Ed Wilson at 17, on the day of arrival at the University of Alabama, September 1946.

in gardens, and they were hard to miss, their tiny bodies turned by the refraction of sunlight into particles of metallic greenish gold as they zigzagged and spun in circles on leaf surfaces, lifted high like dancers on long legs too slender to be seen by the naked eye. The Dolichopodidae are not your usual flies—in the common perception, repulsive seekers of filth and carcasses—but rather, shining clean predators of other insects their size and smaller.

I learned by reading that over five thousand dolichopodid species were known by science from around the world, with many more waiting to be discovered and described. Their biology remained largely unstudied. Here, it seemed, was a subject worthy of lifetime scientific study, awaiting specialists to understand and make it known. I could picture myself in future years: Edward O. Wilson, expert on the dolichopodid flies, position as dipterist in the Smithsonian Museum, expeditions planned with net and bottle to the Amazon, to Patagonia, to the Congo . . .

But then, I discovered something even more exciting. Or, providentially, they discovered me. Ants, as I will next explain.

4

ARMY ANTS

THE COLONY POURED out of a hidden bivouac into our backyard, a horde the size of a dozen Roman legions, forming a line three or four abreast. They were army ant workers and soldiers: upward of one hundred thousand strong, accompanied by their thimble-sized mother queen, on the move hard and fast from an old stronghold to a new. Each ant ran double-time, following a chemical trail laid by scouts ahead, hemmed in by her sisters front, back, and to the side. The whole colony emerged from the bivouac into the yard like the uncoiling of a rope.

Walking along the marching column, I came to the end, and another surprise: The rear guard consisted not of ants but of small beetles and silverfish. These camp followers, which, in time I found, characterize army ant colonies of all kinds, are social parasites. Eluding the jaws and stings of their hosts, they scrounge whatever bits of food they can.

The marching colony, I later learned, was of a kind

often referred to as one of the "miniature army ants" belonging to the genus *Neivamyrmex*. Decatur was close to the northern limit of army ants of any kind.

I met the species again as a freshman at the University of Alabama, when in the woods at nearby Hurricane Creek I discovered army bivouacs in rotting pine logs. I could scoop up entire colonies in this passive part of their life cycle and transport them to the laboratory for close study. I was strictly forbidden from letting them march through the halls of Josiah Nott Hall, the headquarters of the department of biology, but I could study them in their quiescent state. In so doing, I made a remarkable

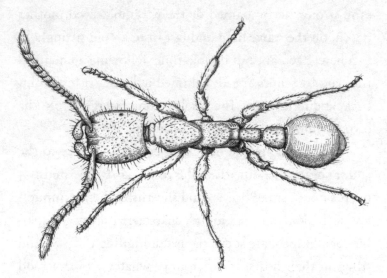

A worker of the army ant *Neivamyrmex*, found in the United States as far north as the Tennessee River. (*Drawing by Kristen Orr.*)

discovery. I found none of the silverfish seen in the Decatur procession, but instead an abundance of tiny beetles, among the smallest of any kind on Earth. They were, I learned later, members of the genus *Paralimulodes*, the first ever found outside South America. On stiff, short legs they skittered over the bodies of the ants and hopped from one to the next. They behaved like fleas. And what did they eat? Not what you'd expect. They licked up oily liquid from the body surfaces of their relatively gigantic hosts. The ants seemed not to mind their ministrations. They made no effort to catch or chase them away.

Years later, during a field trip to Louisiana, as I slept on an air mattress laid on the forest floor, I encountered army ants in a wholly different way. Sometime in the middle of the night I awoke to find ants swarming over my mattress and onto my body. They were the same *Neivamyrmex*, or a close species, likely on a march to a new bivouac. For them my body was just another obstacle to maneuver, like a backyard fence at home in Alabama.

The legionary behavior of army ants is understandably bizarre to humans, but during the grand million-year drama of ant evolution, it has been an outstanding success. Every continent save Europe and Antarctica has its own genus (group of related species) or multiple genera of legionary ants. *Neivamyrmex* is the example in North America and on south into the tropics. The ferocious *Eciton* are the inspiration of the famous 1938 short story by

Carl Stephenson, "Leiningen versus the Ants," and the subject in turn of the 1954 film *The Naked Jungle*, featuring Charlton Heston with Eleanor Parker bravely at his side defending his cocoa plantation from a mile-wide mass of biting, stinging ecitonines. And far from least among army ants are the *Anomma* and *Dorylus* driver ants of Africa, which in real life rival or exceed even the fiercest ecitonines.

The ecitonines, confined in most cases to tangled understory vegetation, move frequently from site to site, ready to attack and fight to the death any intruder of their bivouacs or foraging legions. Many among them form a soldier caste armed with long, scimitar-shaped mandibles, making the colonies even more difficult to study. However, just that was accomplished with high distinction by a psychologist turned entomologist, Theodore C. Schneirla, who conducted research on ecitonine army ants primarily in the field from 1933 through 1965. His conclusions were confirmed and extended during the 1960s by his equally brilliant acolyte Carl W. Rettenmeyer. Their work, with that of others, was summarized in my own synthesis, *The Insect Societies*, published in 1971.

As a young scientist, I knew Ted Schneirla personally and followed his research closely. I found him a calm, intense man, deeply serious about his work. He was guided by two overarching goals linked in purpose. The first was to thoroughly understand these intricately organized social insects, firmly driven by instinct. Next, being a psychologist, he wanted to show that their repertoires of behavior are guided by individual learning. If well-defined instinct in a small-brained insect is a product of learning, Schneirla reasoned, then so are all other forms of behavior. A heavy stress on experience and learning was politically very popular during the time from the 1920s to the 1960s, because it contradicted eugenics and offered hope to those proposing individualistic democracy. I don't think, however, that the amazing account of army ant biology by Schneirla and Rettenmeyer was affected by ideology. They described army ants as they saw them.

Some of the best work of the two entomologists was on a particular species, *Eciton burchelli*. It has a mode of hunting unusual among army ants in general, called swarm raiding. As the workers leave the tightly packed swarm, they spread out in a fan-shaped mob that grows into a broad advancing front. Later they pull back, shrink their fan, and return to the bivouac. The *Eciton burchelli* swarm is an overwhelmingly powerful force. A majority of the workers pour out of the bivouac in a living mass of between 150,000 and 700,000 individuals. The fan they form spreads forward at a speed of up to 20 meters an hour. When they encounter a stream or a deep crevice, the forward workers link legs and jaws to form living bridges.

The swarm raiders on the march are horrific in their activities, on a smaller scale worthy of the aforementioned "Leiningen versus the Ants." "The huge sorties of *burchelli*," Schneirla wrote, "bring disaster to practically all animal life that lies in their path and fails to escape." He continued:

> Their normal bag includes tarantulas, scorpions, beetles, roaches, grasshoppers and the adults and broods of other ants and many forest insects; few evade the dragnet. I have seen snakes, lizards, and nestling birds killed on various occasions; undoubtedly a larger vertebrate which, because of injury or for some other reason, could not run off, would be killed by stinging or asphyxiation.

The *burchelli* swarm raids can be viewed as an ecological scythe that cuts back and forth across the rain forest floor. On Panama's Barro Colorado Island, which has an area of about 16 square kilometers, entomologists have found about fifty colonies active at any one time, each for half the day traveling up to 200 meters. They can be heard at a distance, first the rustling and hissing of their footfalls and those of their fleeing prey, then the buzz from clouds of parasitic flies above them. Finally are heard calls of up to ten species of antbirds that take a share of the fleeing prey. The number and diversity of insects, spiders, and other invertebrate animals drops precipitously in the path of a swarm, but still perturbs too small a fraction of Barro Colorado to have an island-wide effect. Think of the army ant colonies not as vacuum cleaners but as the equivalent of fifty large carnivores, for example jaguars or pumas, feeding not on deer and pec-caries but on prey finely divided into creatures of small size and immense variety.

A hallmark of ecosystems is the presence of primary food producers. In this category are the ants themselves, plus a wide variety of other organisms dependent upon them. In my first experience with army ants, the min-iature species of *Neivamyrmex* I found in Alabama, I discovered a paltry crew of such hangers-on, a silverfish and some unidentified beetles. In the army ant colo-nies of the American tropics, on the other hand, Carl Rettenmeyer and later researchers discovered hundreds of species of such guests, including oonopid spiders; circocyllibanid, coxequesomid, laelaptid, planodiscid, scutacarid, macrochelid, neoparasitid, and pyemotid mites; nicoletiid silverfish; carabid, limulodid, staphyli-nid, and histerid beetles; phorid, conopid, and tachinid flies; and diapriid wasps.

Rettenmeyer's list is certain to prove incomplete. Yet even the great variety of these parasites and preda-tors known to date is not nearly as impressive as the tech-niques they have evolved over millions of years to achieve intimate life with their army ant hosts. The limulodid beetles and nicoletiid silverfish clamber over the bodies of the ants, feeding on their body secretions and stealing food the ants bring into their bivouacs. The circocylliba-nid mites ride on the inner curved surfaces of the soldier's long mandibles. Other mites, in the genus *Antenneque-soma*, which are shaped like clothespins, stay permanently clasped to the base of the workers' antennae. Adults of the histerid beetle *Euxenister* ride ant workers like jockeys on horseback, with their long legs clasped around the mid-bodies of their hosts. Perhaps the most astonishing of all is a *Macrocheles* mite that attaches itself to the tip of the hind foot of the worker, from which it siphons blood but also serves as an extra "foot," thus managing not to inter-rupt the running of its host.

The strange world of the army ants and their symbi-otic guests remind us of an important principle of parasite biology—practiced by creatures ranging from disease-causing bacteria to human criminals—that the most suc-cessful parasite is the one that causes the least damage.

5

FIRE ANTS

ONE DAY, AS I sat on a field chair in the center of Dauphin Island, the main Gulf of Mexico barrier island of Alabama, I was seized by a reckless impulse. A mound nest of imported fire ants was at my feet, and I was talking about them on camera for a television special, *Lord of the Ants*. I wondered, as I had many times before, why exactly are these insects called "fire" ants? I had been stung, as have a majority of people who spend much time outdoors within reach of this notorious pest. Usually, however, the attackers are brushed off quickly and the pain is local and temporary.

But I knew these ants can kill you. So, Rule Number One: never sit, stand or fall into a fire ant mound. If you've acquired an allergy to the venom, you might suffer the consequences of anaphylactic shock. If you are with a small child who stumbles into a mound, triggering a massive attack, the result also can be life-threatening.

So I had this impulse: with the camera running,

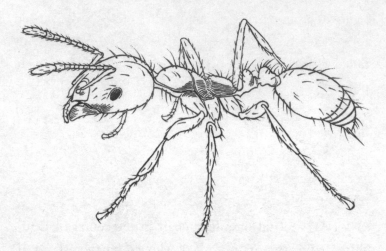

A worker of the imported fire ant *Solenopsis invicta*, a native of
temperate South American wetlands, which was accidentally
introduced into the port of Mobile, Alabama, and spread as a major
pest around other parts of the world. *(Drawing by Kristen Orr.)*

a record potentially permanent, why not experience a
massive attack—then, of course, end it quickly. I would
be able to report definitively why *Solenopsis invicta* is
called a fire ant. Without thinking about it too long,
I thrust my left hand (left because I'm right-handed)
all the way to my wrist into the center of the mound
and held it in place for about five seconds, then pulled
it out and brushed off the large number of ants already
stinging it.

Even in that short time, my hand became densely
sprinkled with ants doubled up and stinging my skin. A
smaller number were running up my forearm in a frenzy

to get to other parts of my body. The colony had not been forewarned, yet its ferocious response was almost instantaneous. In a life-or-death drama for them, fire ants were quicker than the enemy.

The pain was immediate and unbearable. As I described it on the spot to my companions, it was as though I had poured kerosene on my hand and lit it. Within seconds fifty-four defending ants had stung my hand and wrist. I can be sure of that number, because each fire ant sting develops a pustule, and if one scratches the burning sting, it may become infected. The ants seem

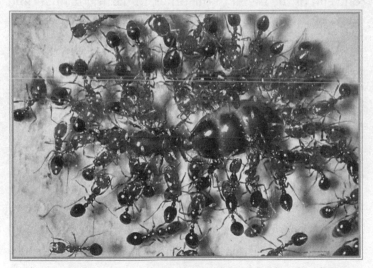

An imported fire ant queen surrounded by her worker daughters, who may come to number in the hundreds of thousands. The workers are attracted by her powerful pheromones. *(Photograph by Walter R. Tschinkel.)*

to say, you have been left with a little reminder: don't mess with our home.

Another notable event occurred the same day. I had brought the film crew to Dauphin Island on the promise that the island—famous as a destination of birds migrating north from Yucatán across the Caribbean—is also notable for the density of fire ant mounds. I believed that there would be many places providing a suitable background against which we could observe and talk about the habits and social behavior of this formidable insect.

But with cameras ready to roll, at first we found nothing to film. The crew and I searched unsuccessfully from one end of the island to the other, in natural habitats and in residents' yards and commercial buildings. Finally, two nests were located within the bird sanctuary, one of which was used in my demonstration of the fire ant stinging defense. How could an entire population vanish overnight? It was as though a giant hand had swept all the colonies away. And that is almost exactly what happened.

I knew the reason. As water rises around a nest of fire ants, or enters the lowermost chambers of the nest, the whole colony unites. The workers come together in a single mass near the entrance. The queen walks or is nudged and pulled into the mass. The helpless young, eggs, larvae, and pupae are carried in and placed there with her. As the water rises to ground level, the clustered colony becomes a raft, ready to float downstream. The living raft thus begins a journey to higher land, in search of a place where the workers can build a fresh domed nest.

The procedure is based on primeval instinct. When the raft touches and stops at any stationary object above the floodwater level, whether a tree branch, a snagged log, or (most hopefully for survival) a rise of dry land, scouts run onto it to investigate. If the landfall is promising, more scouts are recruited. When signs stay positive, the number of workers going ashore rises, they move the queen and young over, and the refugees build a completely new nest around the whole.

I once traveled to Birmingham on a train that crossed the Coosa River in flood. Because water had reached the edge of the railbed, the train moved slowly and at one point stopped, allowing me to peer out a long time in all directions. The broad floodwater plain was peppered everywhere with hundreds of fire ant rafts, all floating slowly toward and around the train, then on away downstream. They were an immense army of refugees in search of a new home.

To complete the story, what happened to most of the Dauphin Island fire ants? Where did all those colonies go? The day before the film crew and I arrived, ten inches of rain had fallen on the island. Torrential rainfall of this magnitude is not unusual for this part of America. Mobile, Alabama, and nearby Panama City, Florida, compete with Highlands, North Carolina, for being soaked with the highest rainfall of any municipal area in North America. The sun was shining when we arrived on Dauphin Island, but an inch or two of rainwater still covered much of the land. The surplus had begun to seep back into the ground. Early on, at the height of the storm, a large part of it had drained northward, into Mobile Bay, whose brackish waters empty farther east into the Gulf of Mexico. And that is what I believe happened to the Dauphin Island fire ants, emigrating on their rafts made by joining their own bodies. To use a famous line from the great gangster movie *The Godfather*, I believe they sleep with the fishes.

6

HOW FIRE ANTS MADE
ENVIRONMENTAL HISTORY

DURING THE SUMMER of 1942, four species of ants lived in the vacant lot next to our century-old family house on Charleston Street in Mobile, Alabama. Exactly four. I know that number with certainty because I examined every cubic foot of the grubby abandoned space, ground, weeds, and rubbish. Working with a sweep net and crawling on my hands and knees I learned every bit of it—as well as the bedroom in which I slept and the kitchen in which I ate my meals. To this day I remember vividly the location of every ant colony that lived in the vacant lot. I also learned a bit about their colony size and behavior. Today I can give you their scientific names.

In 1942 I was an ambitious thirteen-year-old Boy Scout preparing myself, as I imagined it, to someday lead grown-up expeditions to faraway jungles. I devised in the vacant lot what today is called an ATBI, for All Taxa Biodiversity Inventory. The simple but often difficult

Ed Wilson at 13, in summer 1942, with sweep net in the vacant lot adjacent to the old Wilson house in Mobile, Alabama, next to a nest of imported fire ants, the first recorded in the United States.

procedure is to identify all the species of a selected group of organisms in a designated space and time.

This summer in hot, humid Mobile, the fauna of my designated group were the ants. Those I found in the vacant lot proved providential, far beyond what I could possibly imagine at the time.

My first vacant-lot species, I now know, was a colony of *Odontomachus brunneus*, a big, black ant with long snapping jaws and a painful sting. I found a colony nesting in a mix of soil and discarded roof shingles piled beneath a fig tree. There was a colony of the small yellow *Pheidole floridana* beneath an empty whisky bottle. It lived there for almost a year, and I could see it by carefully lifting up the bottle. Next came number three: in a decayed part of the fence that bordered our front yard was a pest familiar throughout the South, the "Argentine ant," *Linepithema humile*. In warm weather long columns of this species foraged out into the lot.

Then came, it can be said without exaggeration, the find of a lifetime—or at least of a boyhood. I found a nest of excavated earth in the form of a mound a foot high, teeming with ants of a kind I had never seen anywhere else. It turned out to be the imported fire ant *Solenopsis invicta*, the first record in the northern hemisphere. For American history as a whole, it was a species of destiny. The species name *invicta*, assigned later by taxonomists, means "unconquered." It was appropriate for one of the

Linepithema humile

Odontomachus brunneus

Solenopsis invicta

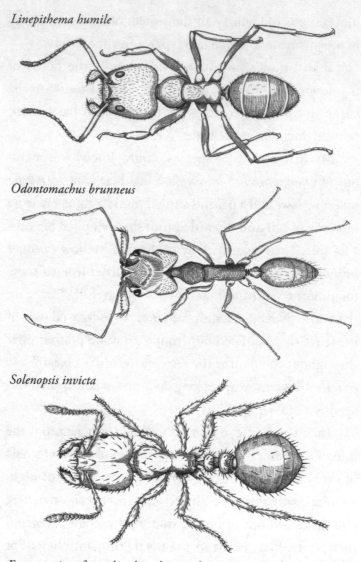

Four species of ants lived in the weed-grown vacant lot next to the author's Mobile, Alabama, home in 1942. One, not shown, was *Pheidole floridana*, a colony of which lived beneath a discarded whisky bottle. *(Drawings by Kristen Orr.)*

most successful invasive organisms of all time. And the word "invasive" is also appropriate. By decree of the U.S. Departments of Agriculture and Interior it means not just "exotic," "introduced," "non-native" or "alien," but also harmful in some way to the environment, or to humans, or to both.

Our house, built by my great-grandfather, an early furniture merchant, was an excellent place to look for newly invasive species. It was within five city blocks of Mobile's commercial docks. Much of the cargo came from Argentina and Uruguay, part of the homeland of the imported fire ant. My father, as a teenage seaman, had made the round trip from Mobile to Montevideo.

My vacant-lot fire ant colony nevertheless could not have been the very first on shore. Had I searched away from the Charleston Street vacant lot during the summer of 1942, I almost certainly would have discovered other colonies in the dock area, and possibly in other parts of Mobile. It seems likely, most experts now agree, that the introduction of the imported fire ant occurred sometime in the 1930s. But not earlier: established colonies grow in size swiftly. They begin to produce and disperse new queens, hence new colonies, within one or two years.

During the remainder of the 1940s entomologists, now aware of the species, watched its populations grow explosively. It filled Mobile and then all the land beyond the city.

Soon what had begun in Mobile became a national problem, then international. The imported fire ant spread to the Carolinas, next to Texas and California. It put ashore on Hawaii and made beachhead in Australia, New Zealand, and China. It also spread south onto several islands of the Lesser Antilles—as though headed back home, island by island. In Alabama it filled lawns, roadsides, and farmlands, reaching up to fifty mound nests per acre, each teeming with as many as two hundred thousand workers, almost all seeming poised to attack intruders. On farms in the surrounding counties, the ants consumed seedlings of radish, alfalfa, and other money crops. They rendered pastures used for cattle difficult to attend. They managed to forage into rural homes, stinging freely.

It soon was discovered that in a few natural habitats, among them especially open pineland, the imported fire ants were attacking small mammals and ground-nesting birds.

Later, as a nineteen-year-old senior at the University of Alabama, having become locally famous as an expert on ants, I was asked by the Alabama Department of Conservation to study the rapidly expanding populations, map their spread, and assay the harm they were causing.

Because the foot-high mounds are so conspicuous and because the impact on people was so striking and widespread, and because I was assisted by James H. Eads, a fellow student who providentially owned a car, the survey proceeded rapidly.

At the price of countless stings and pustules, we confirmed the nature of the damage caused by the invaders. We also picked up considerable new information about the fire ant life cycle. Of importance was the discovery that a single newly inseminated queen can fly up to five miles, build a little nest, and rear worker offspring fast enough to create new queens within two years.

In short, we discovered that the imported fire ant would be very difficult to defeat by conventional means—and in particular by the use of insecticides. This reservation, however, did not stop the U.S. Department of Agriculture and the chemical industry planning precisely that: spray the entire range of the imported fire ant with pesticides in order to wipe it out in one strike.

At first, the ambition seemed justified. The 1950s were a period of American triumphalism. We had saved the world from the armies of fascism. We had slowed if not halted global communism. Our science and our technology were achieving wonders. Americans were ready to think big, really big. We could do anything. Having graduated from atomic to hydrogen bomb weaponry, it was natural to think of the peaceful uses possible in nuclear detonations. In 1957 the Atomic Energy Commission prepared to take a literal, not merely rhetorical, giant step: use nuclear explosions as giant shovels. We could free hitherto inaccessible natural gas deposits. We could, if we wished, scoop out a new harbor in Alaska. And best of all, we could dig a new channel parallel to the overcrowded Panama Canal, achieved with a string of nuclear blasts that would join the waters of the Pacific and the Caribbean. All this portended profound effects on the environment— none good.

Each of the megaproposals soon met with predictions of geological disaster, and they died away. But their spirit did not, and it soared on up with America's triumphs in space, medicine, and basic science. The same spirit also made it seem natural that a major invasive insect such as the imported fire ant could be halted, if not eradicated, by an American force majeure.

In 1958, the U.S. Department of Agriculture planned to engineer the spraying of most of the infested area of the southern United States with the pesticides heptachlor and dieldrin. The fire ant populations would be greatly reduced, but far from completely erased. At the same time, the quantity of wildlife, including mammals and birds and with certainty other insects and invertebrates as well, would also be diminished. Finally, people in the treated area were also put at risk: heptachlor can cause liver damage and dieldrin is a neurotoxin.

The entire effort at mass control was critically flawed because of another, insoluble difficulty: if, despite soaking the terrain with pesticides, a single fire ant colony was left alive, it would proceed to do what every colony does with genius: produce hundreds of winged queens, each able to fly out five miles or more and establish a new colony. This singular biological feat was the reason that I later called the mass spraying "The Vietnam of entomology."

At about this time Rachel Carson entered the fire ant scene. She was appalled at what America was doing to itself. Because I was at that time known as an expert on fire ants, Carson wrote to me with the suggestion that she come down from her summer home in Maine to Harvard for a discussion on the overall problem. Then she canceled, reporting that she had become ill. I followed up by recommending a recently published technical work on the effects of wide pesticide application. I believe that helped, but I have always regretted not dropping everything and driving up to Maine to meet this great American personally.

Rachel Carson, however, didn't need further help. In 1963 she published *Silent Spring*, which revolutionized our thinking about pesticides, and, more than any other single event or contribution, launched the new era of environmentalism. It is worth noting that one ant species, first seen (to the best of my knowledge) on the edge of a vacant lot in a Southern port city, played a significant role in its appearance.

The story of the imported fire ant leads me to turn back five hundred years, when another fire ant, *Solenopsis geminata*, changed history during the colonization of the New World. Or so I concluded, after a good deal of research that combined history and entomology. This analysis suggests a rule of environmental history: all human-caused disasters repeat themselves. The story follows.

7

ANTS DEFEAT THE
CONQUISTADORS

FOR HALF A millennium, an entomological mystery had
shrouded the early history of the New World. During or
soon after 1518–19, a plague of stinging ants struck the
fledgling Spanish settlements on the island of Hispaniola.
According to the eyewitness account by the colonial histo-
rian Bartolomé de Las Casas, the ants destroyed or made
inaccessible a substantial part of the early crops, and they
infested the first dwellings.*

The colonists found themselves powerless against the
rapidly expanding swarms, which also spread to their
Cuban and Jamaican settlements. In desperation, they
designated a patron saint in their plea for divine help.
They conducted a religious progression through the little
village of Santo Domingo. Finally, they moved the settlement across the Santo Domingo River.

* The account here is adapted from my documented article, "Early ant
plagues in the New World," *Nature* 433 (7021):32 (2005).

It was all to no avail. The colonists talked of leaving the island altogether.

Fortunately for modern science, some of the key traits of the Hispaniola plague were reported by Las Casas. The ants, he said, were very aggressive, and able to deliver painful stings. They occurred in dense populations around the root systems of shrubs and trees, yet did not attack aboveground vegetation, as do the ubiquitous leafcutter ants. They nevertheless somehow still damaged roots. And, finally, they became formidable house pests.

As time passed, over a period for which we have few records, the plague subsided. Nearly four hundred years later, in 2004, while I was in charge of Harvard University's ant collection, the largest and most complete in the world, I decided to try to identify the plague ant and deduce, if I could, why it had so dramatically expanded and then subsided. Like a veteran of a lost war, the Dominican invader might still survive, unnoticed, among its more pacific fellow species.

I had already been studying the West Indian ants, with collecting trips to Cuba, the Dominican Republic, Puerto Rico, and the Lesser Antilles, from Tobago to Grenada and Barbados. I had a firm idea of which species might be a covert plague ant and which could not.

There are about 310 species of ants known to occur in the West Indies. I decided not to jump to a conclusion on the culprit species, but to work like a detective, eliminating "ants of interest" until I had a solid suspect. I might then remand my choice to a jury made up of my fellow ant biologists.

To make a long story short, and cut through a large amount of data from field studies, the only species present with all the qualities listed by Las Casas is the tropical fire ant, *Solenopsis geminata*. This inference is strengthened by the close similarity of the plague to the one unleashed upon the U.S. Gulf states in the 1940s by the imported fire ant *Solenopsis invicta*.

This second fire ant species, *Solenopsis geminata*, punisher of the Spanish colonists, is distinguished by a soldier caste with even more swollen heads than regular soldiers, along with powerful jaws used to grind seeds. Its native range is evidently the coastal plain from North Carolina through the Gulf states to Mexico. It is a superb colonizer, having traveled with human commerce and settled in Africa, Asia from Taiwan to India, Polynesia, and Australia. Recent molecular studies have revealed that in the 1500s it hitchhiked on Spanish galleons from Acapulco to Manila and from there on to China.

Solenopsis geminata has at least three traits that make it a superlative "tramp" species. It flourishes on a broad diet that combines prey, dead animals, and seeds. It thrives on beaches, providing its colonies often ready entry. And it travels well in the kind of soil and rocks once carried as ballast in Spanish galleons.

The time line provided by Bartolomé de Las Casas now falls nicely into place, at least for a scientist playing prosecutor. The Hispaniola ant plague struck two to three decades after the arrival of the first Spanish galleons, about the right amount of time for the fire ants to debark and multiply to invasive-level numbers, then subside to the more pacific level they enjoy today as (almost) normal members of the ant fauna.

8

THE FIERCEST ANTS IN
THE WORLD, AND WHY

AMONG THE THOUSANDS of species of ants I have seen during a lifetime of research around the world, I have encountered an immense array of personalities. The least appreciated and most timid, the one with none of the warrior spirit we usually associate with ants, is a small, slender species, identified later as *Dolichoderus imitator*, which I encountered in the Amazonian rain forest north of Manaus. Its colonies consist of a few hundred workers, and presumably the rarely seen mother queen. They nest in random cavities of decaying leaf litter. When I disturbed them, even slightly, they scattered in all directions, picking up as they ran any odd immature nestmate encountered, larva or pupa. Like people fleeing a tornado, they chose as shelter any covered place they encountered, and hid there. I had trouble finding even a few specimens for later taxonomic examination. Had I destroyed the colonies? Almost certainly not. When I lumbered off, a monster one million times the size of one ant, the colony must have reassembled.

Ant colonies possess superb resiliency. Timidity pays off in the rain forest, if you can run fast.

Now to the opposite end of the aggression spectrum. Which are the fiercest ants of the world? And why, in the first place, is there a spectrum among species that spans from pacifist to warmonger? I'll now describe a half dozen of the leading candidates for extreme bellicosity, because together they illustrate a basic principle in the evolution of ants and other social animals—including humans.

The first candidate is to be found among Australia's ninety-four bull ant species comprising the genus *Myrmecia*. The largest, about the size of a hornet, have similar dispositions and deliver shocking stings, also

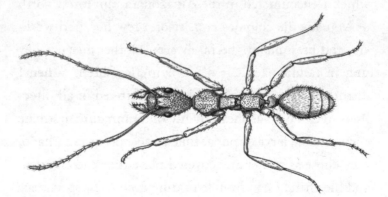

Possibly the least offensive ant in the world, *Dolichoderus imitator* of the Amazon rain forest. *(Drawing by Kristen Orr.)*

hornet-magnitude. They nest in conspicuous craters in the soil, out in the open, with a single entrance surrounded by a ring of excavated soil. They have large eyes and an enduring lack of tolerance for animals the size of humans. Sentries on the nest surface turn and watch as you approach. If you come close, they will start to walk toward you, and it will be a mistake to linger. As you retreat, they will follow you as far as ten meters.

Only the bold and brave would think of walking up to a bull ant nest and start to excavate it. Under normal circumstances, it would be like taking apart a hornet paper nest layer by layer.

However, on an expedition to southwestern Australia in 1955, I learned how to do just that, at a price of no more than a sting or two. I was taught by Caryl Haskins, a fellow ant enthusiast (and, at forty-three, the newly appointed president of the Carnegie Institute in Washington, DC). Haskins knew how to capture entire colonies for laboratory study. First, as he showed me, you approach the bull ant nest while picking up each ant encountered between thumb and forefinger and popping it into a large jar—very quickly, in about three seconds, before it can bring the tip of its abdomen around and sting you. When all the sentries have been eliminated, you proceed to dig into the nest an inch at a time, capturing each ant as it emerges to attack you. At the very bottom, huddled and quiet in the lowest chamber of the nest, you will find the mother queen.

Impressive as are bull ants, due in part to their large size, still more formidable are ants that live in close symbiosis with species of bushes and trees. When the symbiosis is obligate—meaning that neither species can survive without the other—the ants are swift and suicidal in response to any disturbance. To brush against a small rain forest tree of the genus *Triplaris*, for example, is to be assailed so quickly and violently as to create the sensation of brushing a nettle. The guardian ant is *Pseudomyrmex triplar*. Although I've experienced the wrath of these warriors personally, I believe it is more effective to quote José Celestino Mutis of Colombia, the first scientific naturalist of the New World, on an experience he had around 1770. It concerned a small tree in Colombia known as the *palo alto*.

One hot day in the Vega del Guadual, when I suffered from the sun, I paused to stand with my shotgun beneath a somewhat thick-topped pyramidal tree. In a short time, I found myself covered with some red ants, stinging me so fiercely that, with difficulty, I first took off my shoes in order to get undressed, and then took off my shirt, beating my whole body with it [to shake off the ants]. However, since there were so many I had no other choice but to go into the river, shaking my clothes once I pulled them free. Returning to the house, and feeling inflamed all over, I told my story. The slave there

told me that they were ants of *palo santo* [holy stick].
Accompanied by this mulatto I came back to the same
place, where there were many trees of the same species and, in the open fields nearby, many medium-sized
trees [of the same species] one half to three quarters of
a vara [old Spanish unit of length: 1 vara = 0.84 meter]
high, and one vara wide, without any leaf, resembling
a wicker. When you touch the stems with your hand, a
large number of ants pour out through some impercep-
tibly tiny holes, then swarm all over the stem with the
same speed they show in stinging.

Precisely the same relationship has evolved not once,
but in multiple instances, in the rain forests of West and
Central Africa. The ants belong to the genus *Tetraponera*,
and their host plants include vines and small trees. Their
stings are exceptionally powerful, similar to those of the
notorious parasitic wasps called velvet ants. The pain lasts
for hours, and the sting sites often turn into infected pustules. "As soon as any portion of their host plant is disturbed," said the Harvard biologist Joseph Bequaert in
1922, "they rush out in numbers and heartily explore the
trunk, branches, and leaves. Some of the workers usually
also run over the ground about the base of the tree and
attack any intruder, be it animal or man."

The general principle of evolution emerging from the
military strategy of ant colonies is the following: *The more*

defensible the nest site and the more valuable the resources it contains, the more powerful the defense and the greater the fierceness with which it is applied. In short, ants are as mean as they have to be in order to protect their home. No more, no less. Then, in search of the fiercest species, what about the fire ants? Aren't they high on the ferocity list? Yes, but not at the level reached by the plant-dwellers that come out of their nest just to chase passers-by. Only when a fire ant nest itself is invaded, as when a human kicks a mound open with a well-shod foot (as I do almost reflexively out of curiosity whenever I pass by one), do the hordes of stinging workers pour out. If you leave them alone, they will leave you alone.

Nor are the most vicious fighting machines the army ants. Their moving columns and fans drive all creatures before them, but they are selective predators in search of prey, not palace guards. They are comparable to flocks of starlings scratching and pecking their way across an English meadow in search of insects.

Nor, finally, are the mighty colonies of leafcutter ants, several million members strong, whose soldiers ("major workers") attack any enemy invading their nests. Most impressive among them are the supersoldiers (also called supermajors). Equipped with massive jaw muscles that fill swollen head capsules and knife-like sclerotized teeth, they can cut through any soft material.

I've studied the leafcutters in both the field and the

laboratory, where at Harvard University I maintained healthy leafcutter ant colonies in rows of plastic boxes, allowing me and Kathleen Horton, as laboratory organizer, to keep the population far longer than the New York City–magnitude nest found in the wild. Ours was a practice comparable to Japanese bonsai. The small colonies flourished.

The most vicious ants in the world are, in my thoroughly bitten, stung, and formic-acid-sprayed judgment, the epiphyte garden-ants, *Camponotus femoratus*, distant cousins of the large black carpenter ants *Camponotus pennsylvanicus*, which are abundant as forest dwellers and house pests in North America. *Camponotus femoratus* colonies abound in the canopies of the Amazon rain forest. Using soil and miscellaneous vegetable detritus collected from the ground and surrounding branches, they build spherical "ant-gardens" around the epiphytes—species of plants already adapted for growth on the trunks and canopies of the rain forest trees. The nests as a whole are globular in shape, spongy in structure, and held together by the epiphytes that stick out on all sides. In the gardens, feeding on the sap of the epiphytes, are "cattle," scale insects and mealybugs that provide their ant hosts with excrement rich in sugars and amino acids, and sometimes their own bodies when the hosts suffer a shortage of protein.

The ants have been observed carrying seeds in and around the gardens, leading to an early belief that the ants plant them as part of the symbiosis. There are plausible alternative explanations for the phenomenon. The seeds could be collected for food and then misplaced. Or they could be blown into the site during the early construction of the nest.

Whatever the provenance of the Amazon ant-gardens, they and their protein-rich inhabitants are tempting targets for a wide variety of vertebrate and invertebrate enemies. The *Camponotus femoratus* colonies have a reputation among entomologists for a frightening degree of ferocity, but given the location of their gardens (mostly in the rain forest canopy), observations of them in their natural setting are not convenient.

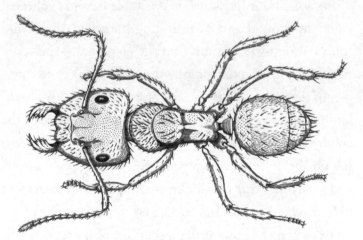

A major worker ("soldier") of the fierce tree-dwelling Amazon ant *Camponotus femoratus*. *(Drawing by Kristen Orr.)*

I finally encountered and was able to study a previously undisturbed garden when I came upon one that was for some reason on a tree branch only two or three meters above the ground. When I turned and walked downwind toward the colony, a swarm of workers erupted almost instantaneously. As I came closer, but still without touching the nest, the defenders went berserk. Piling up on top of one another, they reached out toward me with the abdomens of many pointing in my direction and spraying a cloud of formic acid. Whether their extraordinary fury was triggered by sight or smell I am not sure, but I hereby give them my vote of "most ferocious."

9

THE BENEVOLENT MATRIARCHY

AS I FIRST put pen to paper—literally, I don't type—on a summer day in 2018, 15,438 species of ants in the whole world have been recognized and given a Latinized name. I'll make a wild guess from what I subjectively feel, having described approximately 450 of those known to science, that there may exist 25,000, discovered and waiting to be discovered. The ant taxonomists Stefan Cover and Steven Shattuck, working on the world's largest collection at Harvard University, subjectively put the figure at between 25,000 and 30,000.

The earliest student of ants in the modern scientific era was Carl Linnaeus. He described the first ant species and gave it a two-part Latinized name, cited today formally as *Camponotus herculeanus* Linnaeus 1761, and informally (in English) as the carpenter ant. It is relatively large for an ant and occurs in the cool north temperate zone worldwide.

In 1946, when I started as a student at the University of

Alabama, about two dozen specialists around the world were publishing technical articles on ants. Now there are hundreds, adding to the roster of known species and in ever finer detail their social biology. What the scientists have learned is the achievement by ants, across 150 million years, of almost every adaptation conceivable by a lineage of social insects able to fill the terrestrial world—and to influence, over and again, almost every conceivable niche that contains it.

Ants, using the devices of social organization imagined or in some cases exceeding imagination, have thus achieved among them somewhere within Earth's great formicid queendom species that walk under water to harvest the bodies of drowned insects; other, arboreal forms that glide on body flanges, like flying squirrels, from branch to branch in the forest canopy; still others, huntresses with trapjaws that slam shut in the fastest animal movement ever recorded; foragers in rain forest that find their way home by memorizing patterns of the canopy above them; colonies that enslave captives of other colonies; suicidal soldiers that explode their own bodies by violent abdominal contractions; social parasitic queens that depose the mother queens of victim colonies; hyperparasitic queens that enslave or kill parasitic queens; tiny parasitic queens that ride on the backs of host queens; colonies that unite to form continuous supercolonies that extend for tens of kilometers; gardening species that live on fungi grown on chewed leaves. And much more.

Despite this diversity of trades, reminiscent of a human empire, the majority of ant species have what may be called the ant-standard life cycle of colonies. It begins when a future mother queen, a virgin, leaves her colony of birth and embarks on her nuptial flight. She may mate in midair, or land on a prominent spot, such as a leaf, or twig tip, then release a sex attractant that diffuses into a cloud drifting downwind. Males follow the aerial trail upwind to find the queen. After they meet and mate, the queen searches for a site in which to begin her own colony. The male, on the other hand, having performed the one and only act for which he exists, dies. Within hours he becomes the prey of a bird, spider, or another predator. Or perhaps, if in a city, a desiccated corpse in a pile of others beneath a street lamp.

The queen, guided by instinct, presses on to her destiny. According to species, she searches for a hollow twig, or a space beneath the surface of a rotting log, or a small bare patch of soil free of roving enemy ants in which she can excavate a small burrow with a closed entrance.

The odds are heavily against her further success. In the territorial leafcutter ants and fire ants, and many other kinds, the terrain near any large colony is usually patrolled by scouts and foragers. When this is the case, fewer than one in a thousand of the would-be mother queens lives to bring through her first brood.

The desperation of the pioneers is moderated a bit by

the imported fire ant through the habit of newly inseminated queens to assemble in groups. Up to a dozen or so collaborate in digging and defending the incipient nest. This stratagem, even when successful, still carries a terrible risk for each of the cooperating mothers. Soon after the first brood of workers emerge as six-legged adults, they execute the queens one by one, spread-eagling and stinging them to death until only a lucky single queen is left.

The workers do not spare their own mothers during the paring down. They choose instead the most fecund of the queens, evidently sensed by comparing pheromones among them that signal their varying levels of fertility. They then join in killing their own mothers or allowing them to be killed, if they fail the test.

Colony founding by a queen in most kinds of ants is like a sprint at a track meet. In order to win, you have to get to the correct place at the right time (find a mate, then a predator-free place to start a colony), next be fast off the mark (quickly and safely raise the first batch of workers), and finally reach the finish line (create the largest number of workers in the shortest possible time).

The ant world is almost exclusively a female world. Adult males, with the exception of competing for access to virgin queens, and the food and grooming they receive from their sister workers, are pathetic creatures. They are built for the one-time act of mating—if they can manage even that much. They live almost all their lives inside the

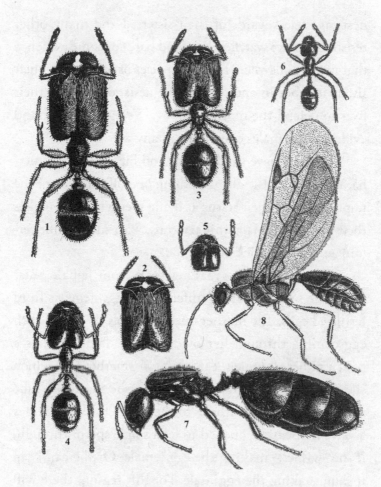

The full complex adult caste system of *Pheidole instabilis*, a species of southwest North America, as illustrated in the classic 1910 drawing by William M. Wheeler. The large queen (bottom, 7) has shed her wings after mating and starting a colony. Above her, a permanently winged male, which dies after leaving the colony to mate. The highly variable worker class ranges from the small workers (6) through intermediate workers (2–5) to grotesquely large-headed soldiers (1).

nests as useless wards of their sisters. I and many other researchers have watched long and hard for some evidence that males somewhere in some species or other assist their sisters in labor in and around the nests, or else risk their lives to defend the mother colony. We have never found evidence that males contribute in any way.

Male ants have small brains and big eyes and genitalia, required for the all-or-nothing one-off purpose of the nuptial flights. Their status as flying sperm missiles makes their mating flights kamikaze runs, with success not certain and only a quick death guaranteed.

What genetically separates male from female ants? Males are produced from unfertilized eggs, females from fertilized eggs. The mother queen controls the sex of each egg coming through her oviduct to be laid. She has a pouch in her abdomen, called the spermatheca, in which she carries the sperm acquired when she mated. A tube runs from the spermatheca to the oviduct. The tube has a valve that can be opened to let a single sperm through, if she "wishes," making the egg female. Or she can keep it shut, making the egg male. For this reason, there will never be a male king, president, combat officer, or anything else other than a male short-lived royal consort.

10

ANTS TALK WITH
SMELL AND TASTE

SPRING HAD ARRIVED when in 2018 I visited the wildlands of Florida's Torreya State Park. The hibernation of
animals had ended. Prey abounded for the delectation of
the carnivores. It was time for a copperhead I watched to
regain its place in the serpentine world. It glided into the
midst of the biologists gathered there as smooth as flow-
ing water. It flicked a forked tongue in and out, delicately
touching but not licking microscopic quantities of odorous material, to brush Jacobson's organ, the chemically
sensitive epithelium on the roof of its mouth. This dramatically beautiful animal, having left the winter retreat
of a nearby copperhead hole, was smelling its way to an
unknown place that instinct and circumstances of the
moment would decide.

The copperhead lived, as do a large majority of other
animal species, in a sensory world scientists have barely
begun to understand. Humans are among the rare

exceptions: we rely almost exclusively on sight and sound. Birds, crickets, frogs, coralline fish, and a few other animals are also trapped in an audiovisual bubble, and find it difficult even to imagine the inner world occupied by the ten million species or so that compose the rest of life.

No one has ever written or told a human love story using perfume and underarm musk.

If our two-legged audiovisual primate ancestors had not originated on the African savanna, if they had not then stumbled through six million years of evolution finally to yield *Homo sapiens*, the land environments today might well be dominated by social insects. Ants would rule as predators and termites as digesters of dead vegetation, both communicating almost exclusively by taste and smell.

Put somewhat more technically, the social insects speak and listen mainly by pheromones, chemical substances passed back and forth as messages between individuals of the same species, and by allomones, substances used by other species and employed by social insects to hunt prey or avoid becoming prey.

Among all of the organisms that live by smell and taste, ants are the virtuosos of chemical communication. If advanced forms of life are found on other planets (and they will be eventually, at least on star systems other than the sun), the aliens will most likely include eusocial species of some kind, by definition those in which societies are formed by means of altruism and advanced degrees of cooperation. Eusocial colonies comprise "royal" members that reproduce with the aid of workers less often or not at all. The workers, freed partly or entirely from the anatomical and physiological necessities of reproduction, are able to specialize and compete more effectively, enabling their colony to outcompete individuals and other colonies. Evolution then proceeds by natural selection, not just among individual members that make up colonies but also among the colonies themselves. The process remains consistent with the fundamental rule of evolution by natural selection: the unit of evolution is the gene, and the target of natural selection is the trait prescribed by the gene. That is all that matters in fundamental Darwinian biology, likely to be a law throughout the universe.

Eusocial colonies have arisen at least seventeen times in the history of life on Earth. Eusociality has emerged three times in shrimp of the family Alphaeidae, known commonly as pistol or snapping shrimp, and notable for creating nests by excavating burrows in the living sponges of shallow marine water. Two other independent lines that gave rise to eusocial colonies exist in the present-day hornets, yellowjackets, and paper wasps. Two more have been found among bark beetles. Two more again exist in the subterranean naked mole rats of Africa. Still other lines arose to produce eusocial groups in modern thrips, aphids, allodapine bees, and augochlorine bees. Finally, a plausible case can be made for the inclusion of humanity, where grandmothering, extreme labor specialization, active military service, and abstemious religious sects are common.

How many ants coexist with humans today? If we make a very rough estimate of 10^{16}, ten thousand trillion, which to the power of ten is about one million times the number of human beings alive at the same time, since a typical ant weighs between one and ten milligrams, one-millionth the weight of a human, the remarkable result of these several guesstimates together is that all the living ants weigh about the same as all the living humans.

However, neither anthropologists nor entomologists should be overly impressed by the evidence of even a very rough biomass equivalence between ants and people. If, in imagination, every one of the 7.5 billion living people were log-stacked, the whole of humanity would fill a cubic mile of space, a mass that could be hidden in a remote section of the Grand Canyon.

For more than 100 million years, ants have played a major role in creating the world we inhabit today. They have done so while creating their own, unique chemosensory world. In other words, they run their complex organizations almost entirely by taste and smell. To identify the chemical signals passed from ant to ant, and by experimentation "speak" human to ant, would be to decipher the Rosetta Stone of their existence.

The evidence of the dependence of ants on pheromones

becomes obvious from even a casual contact with colonies of these insects. Consider making the following experiment yourself. Find an ant colony. Invaders in your kitchen will serve. If none is present, go outside on a warm day and look for nests excavated in bare patches of soil, or under a warm rock. Put a drop of sugar water as near the nest as possible, and watch what happens when scouts roaming in the area find it. The first among them to encounter the treasure usually drinks its fill. Then it runs home to the nest, typically along a straight path. (The scout knows exactly where the nest is even if you don't.) The scout regurgitates sugar water droplets to some of the ants in the nest, and then heads back out. Some of the other workers follow her out of the nest. They track exactly the route taken by the first incoming scout, true to every major twist and turn. It is obvious that the ant has laid down a chemical trail for its nestmates to follow. By also regurgitating some of the sugar water to other workers, it is pheromonally shouting, *I've found food! This food I'm carrying! Follow my trail and find more!*

Next, there is a way to reveal an entire suite of pheromones vital to the integrity of the colony. Pick up a worker from next to a nest entrance, very gently, such as guiding it into a bottle. Then after it has calmed down, allow it to rejoin its nestmates. It will be moderately excited, but its nestmates will be much less so. Now repeat the experiment, but capture an ant from another, distant nest.

When placed with this different colony, the residents will attack it furiously.

Wrong body odor! Alien! Alien!

Workers of the second colony have swept the introduced ant with their antennae, which contain the bulk of odor detectors, and immediately recognize it as an intruder. Each worker carries a combination of substances in the oily surface of its body parts specific to its colony of birth and as unique as a lock and key. Those with an alien odor are attacked, killed, or driven away, instantly.

The body odor of an individual ant, a mix of scents absorbed into its body oils, is like the face of a person or the uniform he or she may be wearing. The combination of chemical compounds it carries allows other ants of the same species to tell at an olfactory glance whether it is a member of the same colony or not, its gender, physical caste, approximate age, and the task for which it is specialized at the moment.

Much of this information must be processed in seconds, and it requires unerring precision. The accuracy of multitudes of accurate responses is necessary for survival of the colony, which in membership ranges, according to species, from less than a hundred to over a million.

Watch a stream of worker ants passing back and forth, some moving in one direction and others the opposite. As individuals meet head on, they pause for only a fraction of a second, not hesitating at all, then pass on as close to the same velocity as when they approached. Slow-motion photography reveals the detail in each such exchange. As the ants run along the odor trails laid by the colony scouts, they sweep their antennae from side to side. The first segment of each antenna, the scape, is long and bare, functioning as a stem to hold and move the outer, shorter segments, collectively called the funiculus.

The funiculus is the nose of the ant. It is densely covered with a layer of tiny hairs, knobs, and plates specialized to detect various chemical substances and transmit neurologically their identity and amount to the brain. The ant chooses her action quickly and decisively, through instinct and circumstance, obedient to the sensory information provided.

We still have little understanding as to how the sensory receptors of ants work. At present we largely know only what we see, using our dominant senses of sight and sound. Imaging the surface of one tiny part of the funiculus reveals an astonishing number and diversity of minute structures. While it would be interesting to learn the details of how each such structure functions, the means by which they differentiate fine chemical signatures and fire the nervous system cells to and into the relatively rudimentary ant brain, I must be satisfied to know that I was able to at least be the first to piece together how pheromones were being used as a means of communicating information between ants and within a colony.

11

HOW WE BROKE THE
PHEROMONE CODE

TO UNDERSTAND THE natural history of ants it is important to find a way to communicate with them, no less than if they were humans from another culture, or even aliens from another planet. How could we possibly ignore them? The ten thousand trillion alive at any time share the land mass of Earth with fewer than eight billion human beings. Their lineage, 150 million years back into the Age of Reptiles, is ten times longer than ours.

We nevertheless have a lot in common. Watching the activity of an ant colony is like studying Ambroglio Lorenzetti's *Allegory of a Good Government*, which portrays Siena as a city-state run in perfect order. In this classic panorama, rendered in 1340, the citizens are all productively active and respectfully well-behaved. They bustle about calmly in and out of the streets and buildings and through the guarded gate out into the bountiful countryside. Each citizen is guided by a purpose. For the moment Lorenzetti's city is at peace, free of external enemies. The armies of rival kingdoms, duchies, and city-states exist, but are parked peacefully somewhere beyond the mountainous horizon.

Similarities exist, but real differences between ants and men are profound. Ants create civilizations by instinct—because they are capable of doing nothing other than what they evolved to create. For their part, human beings are torn by the competing needs of self, family, and tribe. We use culture to banish instinct or at least tame it, even while using it to create our values.

Most importantly, as I have stressed, humans communicate by sight and sound, allowing the creation of words with arbitrarily chosen meaning. In other words, we communicate by language, the prerequisite for the most rapid possible evolution of social order. Ants, in contrast, use chemical secretions smelled or tasted with gene-fixed meanings.

Given that the more than seven billion human beings and ten thousand trillion ants jointly rule the terrestrial world, it is of some importance that we are able to talk to them, no less than if we had landed on an exoplanet inhabited by another eusocial species.

These were thoughts that ran though my mind when, in the summer of 1958, I decided to learn the pheromone language of the fire ant colonies I studied in my laboratory at Harvard. That goal may seem at first to be impossibly ambitious, but I had several advantages in approaching it. Most of all, the details of the life cycle and the relative simplicity of a fire ant's life made the species especially favorable for the attempt.

So, how to begin? I used a rule that has inspired my life as a research biologist. *For every problem in biology there exists a species ideal for its solution, and conversely, for every species there exists a problem its study is ideal to solve.*

The ideal problem presented by the imported fire ant was the communal language of foraging for food. Fire ant workers are geniuses among ants in coordinating search and retrieval. Let a colony settle in your garden, and the next thing you know it's stealing cookie crumbs from your kitchen table. To repeat, scouts leave the nest to hunt for food, wherever it may be. When successful, they recruit sister workers to protect and help carry the booty back to the colony. And more: they report to the sister workers the exact location of their prize. When a huntress finds a food particle too large to pick up and carry, a dead mouse, for example, or a piece of cake left carelessly on a picnic table, she feeds on it for awhile, smearing the odor on her mouthparts, making it possible for her nestmates to judge its quality later. She then heads in a relatively straight line back to the mother nest, protruding her sting, dragging the tip over the ground, and laying minute quantities of pheromone.

Reaching the nest, the forager runs inside and turns

to one nestmate after another, exposing them successively to the dinner invitation provided by the trail pheromone, plus the nature and quality of the food discovered. The richer the food, and the hungrier the colony, the more excited the movements of the forager. Translated into humanese, and to repeat, her exhortation, *Food, food, I've found some food* changes to *FOOD! FOOD! PAY ATTENTION. FOLLOW MY TRAIL! I'M GOING BACK!* After dashing from one nestmate to another, she then returns along the trail she has just laid.

It occurred to me immediately that first, the trail contains a pheromone, and second, that the response to it by laboratory experiments is the best bioassay to locate the gland that produces the pheromone.

So I set out to locate the gland. If I could find this part of the ant Rosetta Stone, I could "speak" to a fire ant colony and tell them where to go for food. All I needed was to separate the exocrine glands, those that secrete substances to the outside of the body, and create artificial trails to which laboratory ants might respond, or might not.

Of course, I first needed to know that exocrine glands existed in ants, and where they are. Thanks to the pioneering work in the nineteenth and twentieth centuries by the microscopists Charles Janet, Auguste Forel, William Morton Wheeler, and Mario Pavan, I had a good idea of the location of exocrine glands of ants generally. This earlier anatomical research had used the classical methods of histology, sectioning specimens and reconstructing three-dimensional structures with the use of compound microscopes. Their descriptions helped a lot—in fact, they were necessary—but I still faced a formidable problem. A typical fire ant is only two to five millimeters long and one or two milligrams in weight. How could I remove individual glands, wash them free of contaminants, then present their contents to the laboratory colonies, when they are scarcely larger than a speck of dust?

Work this fine is usually accomplished by biologists with micromanipulators, which can handle extremely small quantities of material with fingertip touch translated through sufficiently small instruments. But I was too impatient to learn a new technology, and as it turned out I was also lucky. I reached the objective with fingertip touch by directly employing the finest of all hand-held instruments, the needle-tipped Dumont Number 5 forceps, used by jewelers to handle very small precious stones. The movements I employed to separate the glands were not consciously muscle-controlled. They were the normal tremors of the fingers made minutely visible under a dissecting microscope. These involuntary movements were just enough to cut through the base of the glands and ease them into insect-blood saline for further preparation.

It was in the course of this exploration of ant pheromones that I came to appreciate another working principle of biological research:

When searching for a new phenomenon, try serendipity. Use precise but rough and easily repeated experiments to obtain some result or other, whether expected or not. The primary goal is to find previously unknown phenomena. There follows the repetition of experiments, careful measurements, search for other ways to test a result, and the construction and testing of opposing explanations, all that must be achieved before committing to a printed report or other formal announcement.

The best result of serendipity is surprise, which I experienced next in my search for the trail pheromone. With my tools and techniques working, I first turned to the venom gland of the fire ant. Foraging workers laying trails can be seen with a stage microscope to extend the sting in order to apply the trail pheromone. It made sense to suppose that the venom came from some gland or other. The pheromone might prove to be the venom itself. But when I tested this hypothesis, the result was wholly negative. Venom meant nothing for the hungry fire ants. Nor did, less surprisingly, the contents of glands I tested from other locations throughout the body.

Then I noticed another candidate organ connected to the sting, the Dufour's gland, named for Léon Jean Marie Dufour, who first described it in 1841. Found widely in ants and wasps, Dufour's gland opens into the base of the sting. Sausage-shaped, barely visible to the naked eye as no more than a tiny white speck, it seemed an unlikely candidate for so important a role as the fire ant odor trail. Yet when I teased one loose from a freshly killed ant, washed it in saline, crushed it against the tip of a sharpened applicator stick, and with it drew an artificial trail away from a nest entrance in the laboratory, the result was stunning. The ants in the nest didn't just follow the trail in modest numbers, they exploded out in a crowded column. They streamed back and forth all along the trail I had laid. They were like people reacting to a very loud fire alarm in a crowded building, running out and about and shouting.

Contact had been established, a biological response to an active chemical. The next step was to use the bioassay to identify the chemical. If successful, it would be a real breakthrough — a word identified in the ant Rosetta Stone, so to speak. Providentially, organic chemists studying natural products had recently perfected a technique that allows the identification of tiny amounts of mixed organic substances. Gas chromatography allows the separation of the substances in the mix, and makes possible the identification of the components separated and identified by comparison with previously identified components, the role of mass spectrometry.

At this point, I was joined by three chemists at Harvard University familiar with the new methodology in a joint effort to identify the fire ant pheromone. They were James A. McCloskey, then a young professor at

Baylor College of Medicine, Houston, Texas; John H. Law, professor at the University of Chicago, later chairman of the Department of Biochemistry at the University of Arizona; and a graduate student, Christopher T. Walsh, later a distinguished faculty member at the Harvard University Medical School. To learn the identity of the pheromone, which was certain to be an organic compound, we needed a pure sample of it in at least milligram amounts. But that requirement posed another problem: each ant contains only millionths of a gram of the pure substance. To proceed further, we needed to collect the pheromone from thousands of fire ants.

Where could the team obtain such a large sample? As the biologist of the group, I knew the answer. Along the highways and byways of the southeastern United States, mound nests of fire ants occurred at approximately every hundred meters on grassy road strips. Each was home to as many as two hundred thousand workers. And I knew how to harvest a large part of the population from each.

So I led my chemist friends to a highway just west of Jacksonville, Florida, where fire ant nests abounded next to ponds fed by quietly moving freshwater streams. Braving the stings of the fire ant workers, we shoveled masses of soil from the mound nests into the stream and watched as the ants in the soil floated to the surface. They next obeyed the same instinct that saves fire ant colonies from natural flooding: they formed masses that served as living rafts. Under natural conditions, the rafts can float down-stream until the ants reach safe harbor.

By exploiting this communal instinct, the Harvard team collected enough ants, and the products of their bodies, to conduct an ordinary organic chemical analysis. Using artificial trails made from purified material, my collaborators learned that the pheromone is probably a terpenoid.

Then they hit a mysterious setback. The more the pheromone was purified to learn its atomic structure, the weaker it became. This reversal of an expected result may mean that the effect is the result of not one but a mixture of substances. It was up to Robert Vander Meer, heading the fire ant laboratory of the U.S. Department of Agriculture in Gainesville, Florida, to show that this perception is correct. The full effect of the Dufour's gland secretion is that it is a mix of pheromones, variously to excite, to attract, and to lead.

One day I played the role of a formicid Moses, laying down a large amount of the concentrated pheromone next to the entrance of an artificial nest. Amid intense excitement, a large fraction poured out into the open. Alas, I had no promised land to offer them, and after wandering about idly they all gradually filtered back home.

12

SPEAKING FORMIC

AMONG THE MORE than fifteen thousand species of ants classified by entomologists, there exists a babble of tongues, each a Formic,* an array of pheromones used by workers to order their social life. Other biologists and I have discovered in part how to translate their chemical language into the audiovisual languages of human beings.

So, how many pheromones does a worker of a given species use? How many words exist in what I hope I may be permitted to call a "formicese"? My guess is somewhere between ten and twenty, the exact amount according to species. In addition, ants are able to create new messages by varying the amount of the pheromone released.

For example, when a harvester ant (*Pogonomyrmex badius*) trying to collect seeds and other food is confronted by a pack of fire ants, its deadly enemy, it spritzes a fine spray of the alarm substance methyl heptanone from twin

---

* "Formic" is the name Robert Frost used for ant language in his splendid little poem "Departmental."

glands at the junctures of its sawtoothed mandibles. The material is volatile, so it dissipates rapidly to produce an odor easily detected by ants, and people. Unless the ant then runs away, the methyl heptanone vapor forms a hemispheric "active space," within which the pheromone can be smelled, by ant or human. The substance, hence the odor, is greatest at the point of release on the front of the ant's head. It weakens exponentially toward the boundary of the active space.

The result of speech achieved by smell is that when first detected by the "listening" ant at the edge of the active space, the signal is weakest. A weak signal serves to attract the receiver and guide it toward a higher concentration of the pheromone vapor. The methyl heptanone is comparable to a flickering red light; it says, *Attention! Something is wrong. Come here and check it out.* As the ant moves toward the source and into a higher concentration of the pheromone, it becomes excited and begins to run around in search of the trouble. *Help. A nestmate is in a struggle. Run to her, into the higher concentration.* Soon, usually within a few seconds, the follower ant arrives at her distressed nestmate and joins the fray. From one pheromone, the harvester ant has fashioned the equivalent of three words.

Can one ant species actually "read" the pheromone language of another? In some cases they can, and the capacity to do so opens the door to their victimization by social parasitism. One example I discovered in the mountain forests of northern Trinidad involved the arboreal species *Azteca chartifex*. The colonies, which grow to immense size, build massive nests of masticated wood fiber. Tens of thousands of workers stream outward in thick marching columns along the tree branches and down the trunks to forage on the rich plant material and insect populations that inhabit the ground vegetation.

While visiting a nature reserve in the mountains of northern Trinidad, I noticed that a second kind of ant, somewhat larger and differently colored (later identified as *Camponotus apicalis*), was running with the *Azteca*. They were following the *Azteca* trails out of their own nest in the tree down to the feeding area on the ground. In effect, the *Camponotus* were using information stolen from the *Azteca* to appropriate part of the *Azteca* food supply. The *Azteca* tried to catch the *Camponotus*, but the intruders were too strong and fast to be trapped and pinned.

Ants are geniuses of olfaction. Dogs have an almost unlimited capacity to distinguish odors, but no more than ants, which know better what to do with them. Ants have built civilizations upon odors to which their brains are genetically designed to respond. Humans do better with sounds and visual signals—words—given arbitrary meanings that can be pieced together with sentences, yielding thereby a vastly larger potential array of meanings.

Ants, for example, can put together pheromones with

other odors to create "proto-sentences." A foraging worker, having encountered fire ants, rushes into its home nest with the equivalent of shouting *Emergency, danger* by spraying alarm pheromones, then *Enemies* by presenting the odor of *fire ants* it has acquired on its integument during a recent tussle, and then *This way, follow me,* as it turns and runs back along the odor trail it has just laid. I can imagine another pheromone in the air implicit to those who go forth in battle: *The Queen! the Queen! Fight to the death for our Queen!*

Might the ants, as well, during their 150-million-year evolution by thousands of species, have evolved a true language? Can they emit pulses of pheromones varying in frequency or amplitude to create words, as we do so well using sound? The answer, established by models from mathematical physics, is maybe, but highly improbable. Odor pulses are fundamentally different from sound pulses. In generating information—to talk with odor, so to speak—it would be necessary to control the emission and reception over a distance of only a few millimeters.

Ants and other invertebrate animals are too small, and their brains too meager, to inch past the limits of communication which they possess. Even so, social insects in general, and ants in particular, have, somewhere among the thousands of living species, accomplished almost every other innovation with chemical communication that we are able to imagine.

13

ANTS ARE EVERYWHERE (ALMOST)

THE MOST FORBIDDING natural environment in the continental United States, rivaled only by Death Valley in July, is the summit of Mount Washington in midwinter. This highest peak in the northeastern United States, at 6,288.2 ft (1,917 meters), is located in the Presidential Range in New Hampshire. It experiences temperatures that are erratic throughout the year and arctic in winter, with frequent subzero temperatures and hurricane-force winds.

I've made two trips over the years (by automobile) to look for ants that might be found on top, at least on the mildest midsummer days. And I found them. There were colonies able to survive and grow mostly under flat rocks out in the open warmed by the meager sun. There were three species, also found as far north at the Arctic Circle: *Camponotus herculeanus*, *Formica neorufibarbis*, and *Leptothorax muscorum*.

Later, also in midsummer, while attending a literary

conference in Sun Valley, Idaho, I traveled with friends on a ski lift almost all the way to the top of the mountain, beyond the upper edge of the forest. I was again in an arctic environment, a grassy field amidst a scattering of dwarf trees. Searching diligently, turning over sunwarmed rocks and scanning the trunks of the stunted trees, our expeditionary party found one of the arctic species, *Camponotus herculeanus*.

About the same time, I received a letter from a naturalist who reported that he and a friend were planning to hike across Labrador the following summer. Would I like to see any ants they encountered? I accepted enthusiastically. Later, they reported that they had encountered ants only once: a single colony nesting at the base of a stunted tree. It was—no surprise—*Formica neorufibarbis*.

During the past hundred millennia, as humanity spread around the world, humans have mingled with thousands of species of ants settled in natural ecosystems. We have accidentally spread hundreds more via our crops and baggage. In one conspicuous respect, nonetheless, ants have remained relatively helpless. They are very poor oceanic travelers. Prime evidence of this weakness is their near failure to occupy and evolve on the Galápagos Islands. Birds and reptiles, in dramatic contrast, have colonized the archipelago, in most or all cases from South America, with dramatic effect.

Darwin's finches are the iconic example of the adaptive

radiations that have occurred on the Galápagos. A single pair, or group, that managed to fly (or were blown by storms) to at least one of the islands survived and flourished. In time, the pioneer species divided into two or more species, each of which retained the capacity to divide into still more species, which then specialized through evolution for a different niche. The result found by Darwin and other naturalists was an entire fauna spawned by the single progenitor: species with their bills adapted to feeding on insects, plus others with bills enlarged and thickened that enabled diets of seeds of varying hardness, and, most famously, in the "woodpecker finch," a bird that instinctively breaks off twigs and thorns and uses them to dig out insects that burrow in wood. Among reptiles, one species is adapted for conventional existence on the land, while a second dives into the marine water and feeds on the bottom vegetation.

No ant species has multiplied to join this spectacular evolutionary drama. When I visited the Galápagos for my own pilgrimage (on this occasion as a scientific adviser to the government of Ecuador), I found ants everywhere, but they were almost all "tramp" species, having originated variously in Africa, tropical Asia, or the continental Americas, then hitchhiked in cargo and luggage to these islands and to other parts of the world.

To my knowledge, only one species has evolved (without multiplying) in the Galápagos by itself: a carpenter ant given the scientific name *Camponotus williamsi*, after an earlier expedition leader. This ant apparently evolved on the islands, but has not split into multiple species.

Unlike birds, coconuts, and people, ants are especially poor at crossing oceans on their own. When in 1967 Robert W. Taylor and I published our comprehensive monograph, *The Ants of Polynesia*, we found these insects to be everywhere prominent among the native insect faunas, including those on tiny atolls. But they are unusual, we also found, in that nearly half of the species have been introduced into the Pacific by modern human commerce. More precisely, the Pacific islands filled up with ants only during the European settlement of the past four centuries. East of Rotura, Samoa, Tonga and New Zealand, there appear to be no truly native ants at all. We found, in contrast, that introduced tramp species play an increasingly dominant role in the environment as one moves in steps to Tahiti and other Society Islands, then the Tuamotu archipelago, the Marquesas, and finally Hawaii.

The case of Hawaii is of special interest to scientists who study evolution and the environment. Its thirty-six known ant species, representing twenty-one genera, include twenty-nine tramp species spread from elsewhere around Earth's tropical regions. Not one is a native, in the sense of having evolved in the islands. All thirty-six species were, it seems, unintentionally brought to Hawaii by people in their food, woodwork, and clothing. It all started, very likely, with the Polynesians who first touched land with their catamarans two thousand years ago.

The great entomologist Elwood C. Zimmerman, working at the B. P. Bishop Museum in Honolulu, pointed out in 1970 that Hawaii was ant-free, with important environmental consequences. "The Hawaiian Islands," he wrote, "constitute a wonderful natural laboratory where we may observe many evolutionary phenomena in active process in the early stages of operation and often in simplified and clearly defined form uncluttered by many of the masking effects that may be present in many of the environments."

Ants were not the only organisms who failed to colonize in prehistory these most remote islands on their own. The result is a land with disharmonic, or unbalanced, ecosystems. There exist no native mammals, reptiles, or freshwater fish. The native flora of two thousand kinds of higher plants represent only about 275 ancestral immigrant stocks. The insect fauna, comprising about six thousand species, represent among them only about one-third of the orders of insects. "Perhaps only one plant or insect founder," Zimmerman mused, "gained access to Hawaii in each 25,000 to 10,000 years of the subaerial history of the archipelago."

Ants, conquerors of most of the rest of the biological world, did not make it on their own. As a result, we are privileged to witness something of the way Earth's land environment would be today if waves of flying queens and seething colonies of ants had never existed.

Those few insects, birds, and other animals, and plants, able to reach Hawaii during the millennia before the arrival of humans and ants, were free to evolve more quickly than in most archipelagos, and many did. The number of species in relative terms exploded in many groups, including crane flies, damselflies, delphaeid leafhoppers, saldid bugs, *Odynerus* wasps, colletid bees, *Hyposmocoma* moths, and among birds, the fabulously diverse and beautiful drepaniid honeycreepers. In a few of these cases there are more species on Hawaii than in the rest of the world combined. All of this production is now shrinking because of the combined activities of humans and their invasive companions, especially ants.

Across generations, in a period of intense field and laboratory study begun around the middle of the nineteenth century, myrmecologists have labored to find, classify, and describe every aspect of the biology of every one of the more than fifteen thousand species of ants alive on Earth, thereby better to assess their evolution and impact on the planet.

In the course of this study, we have found, as discussed, that ants are weak voyagers, unable to cross major water gaps. They have depended on humans for dispersal to distant islands. However, on arrival in a new land, ants show creativity by occupying among them almost every terrestrial habitat. They penetrate every available nest site, take control of most available food sources, and in doing so create an arthropod hegemony that controls every level of the land from the highest canopy to the lowest root mass.

Myrmecologists, I among them, have often wondered whether ants anywhere occupy caves, at least in small numbers. There are strong reasons to believe it possible. The dark, damp soil and rocks are ideal nest sites. Many caves are deep in guano that can serve as a source of primary energy. The same is true of the underground streams and pools formed by the seepage of groundwater. Dispersal is not a problem. Colonizing ants need only walk or fall into the depths.

Ecologists who study life in caves distinguish two major categories of cave-dwelling organisms. "Troglophiles" are attracted to caves and spend part of their lives in them, but also spend time outside. Cave-dwelling bats immediately come to mind. To explore cave mouths is to encounter a large array of troglophiles prepared to migrate in or out of the depths, simply to make a living in the intermediate conditions close by. Among the most conspicuous of such insect troglophiles are the large carnivorous camel crickets of the genus *Ceuthophilus*. Furnished with body spines and sharp, powerful jaws, they enjoy the protection of the caves during the day and the opportunity to spread out to forage for food during the night.

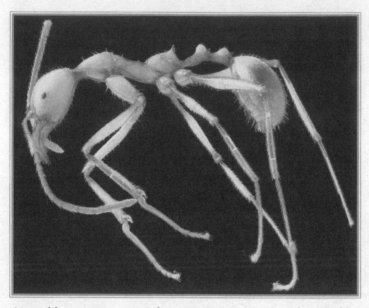

A possibly true cave ant, *Aphaenogaster gamagumayaa*, described by Takeru Naka and Munetoshi Maruyama, from a limestone cave on the island of Okinawa-Jima.

Myrmecologists are especially interested in troglobites, the second kind of cave dwellers, specialized for living permanently deep within the interior, with no trace of light from the outside. And why not? Ants, at least those known to science, are closely suited to dark, moist environments. Multiplying specialized species constantly since the origin of the world fauna over one hundred million years ago, they have enjoyed countless opportunities to settle their colonies and genes into the netherworld.

In 1922, William Morton Wheeler, my predecessor

as professor of entomology and curator of entomology at Harvard University, received a collection of ants made by the ecologist F. M. Urich in Guacharo Cave, today known as Oropouche Cave, Trinidad. The interior was occupied by oil birds (guacharos), the equivalent among birds of the cave-dwelling bats. The ants' pale color, small eyes, and long bristly hairs suggested to Wheeler that they were likely full-blown troglobites, specialists of life deep in caves. He gave them the formal name *Spelaeomyrmex urichi* ("Urich's cave ant"). Today their formal taxonomic name is *Carebara urichi*.

Forty years later, in 1962, I was living with my wife, Irene, on the new Asa Wright's Nature Reserve. I decided—I could not resist—to make my own visit to see Urich's cave ants.

After an automobile trip and a long, difficult walk through forest and cacao plantations, I arrived at the cave entrance. I was tired, tense from mild claustrophobia, yet excited by the prospect of seeing something truly new. The Oropouche Cave is the source of the Oropouche River, which flows from back in the cave as a clear stream several meters in width. Much of the otherwise bare cave floor was covered by guano dropped from the large numbers of guacharos nesting overhead. The insect and other arthropod fauna just inside the cave entrance was rich and diverse, consisting of several species of ants, springtails, cave crickets, earwigs, small flies, and mites. Most, as best I could see, were troglophiles, species that also occur outside the cave entrance.

Fifteen meters in from the entrance, I entered total darkness. The cave continued on through five major twists along a length of 200 to 300 meters. The arthropod fauna declined until it consisted mostly of entomobryid springtails, bristletails, and isopods. Toward the end, the cave ceiling dipped to within a meter of the stream surface, and the passage continued for another twenty meters. Along the way, I encountered several colonies of Urich's cave ant. One I dug up and placed in a container to take back to the laboratory for closer study.

Weeks later, Irene and I moved from Trinidad–Tobago to Paramaribo, the capital of Suriname, then in its last days as a colony of the Netherlands. I commuted to the surrounding tropical forest and grassland to savor the extremely rich ant fauna of the South American continent. At one point, working out from the native village of Bernardsdorp, I began to dissect a large rotting log, which proved a treasure trove of ant and other insect and arachnid diversity. As I pulled away a large piece of decayed bark, I found a colony of Urich's cave ant.

There were no caves nearby. So the species encountered in the Guacharo Cave was a troglophile, a species of the woodland able to live in caves when they are available, and not a troglobite, confined to caves. Ants as a whole could not be said, I thought, to have yet truly colonized this distinctive environment, as they have the driest deserts, the highest canopies, and the coldest ice-free habitats.

Forty years more passed without evidence of any true troglobiotic ant anywhere in the world. Then, two previously unknown candidate species were discovered by separate teams in Asia. Each ant was given a jaw-breaking scientific name: *Leptogenys khammouanensis* from Laos and *Aphaenogaster gamagumayaa* from Okinawa-Jima.

Both of these formicid pioneers have only been found deep within caves. Both have access to nutrients available in abundant bat guano, as well as abundant insects, spiders, and other arthropods that probably serve as their prey. The adaptations they have made are typical of deep-cave dwellers in general: long slender bodies, elongated appendages, reduced eyes, and loss of pigmentation.

High mountain peaks, ice-free subpolar steppes, and caves are the last habitat frontiers for ants. After 150 million years of evolution, ants have acquired a toehold (technically, a tarsus-hold) in caves. More are likely to be found.

14

HOMEWARD BOUND

LARGE RUDDY AND black ants of the species *Cataglyphis bicolor* race on long slender legs across the hottest desert salt pans of North Africa and the north coastlands of the Mediterranean Sea. Some of these foragers obtain food for their colonies by licking edible secretions from surfaces of plants. Others are huntresses, capturing any insect, pillbug, or other creature they can subdue. Even ants of other, weaker kinds are among their common prey. In summer they also become scavengers, picking up the bodies of creatures killed by the heat and desiccation of the scorched ground.

During their forays, the *Cataglyphis* ants travel up to a hundred or more meters from their colony's nest, the equivalent of many kilometers for human beings. The entrance is often no more than a hole in the ground, which expands below into a metropolis of chambers and galleries. Finding their way home, often laden with food for the colony, to one little hole in the desert surface, requires a series of mental operations worthy of a human explorer. As adduced in a lifetime of distinguished studies by the Swiss entomologist Rüdiger Wehner of the University of Zürich and his collaborators, the *Cataglyphis* use an ingenious combination of sequenced instincts and, to humans, impressive topographic learning.

In the opposite of what may seem to be the case, the seemingly featureless celestial hemisphere beneath which the *Cataglyphis* huntresses wander is crowded with geographic information. First, though, they do not have skyscrapers and bridges to set their angle of direction, but there is an abundance of rocks, shrubs, and wallows. Like the better known honeybees, they navigate by the direct light of the sun in a form of dead reckoning using guideposts to which humans are mostly blind, including the spatial gradients of polarized light, spectral composition of light, and the radiant intensity that form cues across the entire vault of the sky.

The wandering *Cataglyphis* worker almost always knows exactly where she is. When she decides to return home, for example when burdened with prey or perceiving the approach of night, she does not spend time climbing a height to look for a landmark, or explore back and forth in search of a previously laid odor trail. She runs in a straight line to the colony nest, and down the entrance.

It sometimes happens that the ant's straight path misses

the entrance. All is not lost, however. The ant knows the amount she has traveled and is aware of her mistake.

With a series of ingenious experiments, Rüdiger Wehner and his colleague Mandyam V. Srinivasan discovered the method by which the ant corrects her mistake. In their words:

If a homing ant (*Cataglyphis bicolor*) gets lost, it does not perform a random walk but adopts a stereotyped search strategy. During its search the ant performs a number of loops of ever-increasing size, starting and ending at the origin and pointing at different azimuthal directions. This strategy ensures that the center area, where the nest is most likely to be, is the one investigated most intensively.

After one hour of continuous search, the ant's search path has covered an area of approximately ten thousand square meters, with the system of loops precisely centered around the origin. In nature, evidently, one seldom encounters a truly lost *Cataglyphis* ant.

Once, before I had learned the homing technique of *Cataglyphis* ants, I used a similar procedure when I found myself lost in the Amazon rain forest. I had been staying at a World Wildlife Fund station north of Manaus. Distracted by the rich variety of species in this unspoiled part of the forest, and by an annoying pet parrot owned by the camp *mateiros* that enjoyed landing on me and sinking its talons into my shoulder, I wandered into the forest depth far enough to realize, suddenly, that I had lost the trail. I couldn't just select a direction at random and travel a straight line in the hope of cutting the trail. That would make the problem far worse. If I chose a line at an angle

A worker of the desert ant *Cataglyphis fortis*, running at high speed with its rear segment tilted upward, foraging along a wavering path away from the nest (N). Finding food (F), it is able to return in a straight line back to the nest. *(From Rüdiger Wehner,* The Desert Navigator, *2020.)*

deviating from the trail, I might end up, days later, sloshing into the wetlands of Venezuela or halting somewhere along faraway banks of the Amazon River. Or worse: sit tight and suffer the humiliation of waiting to be found.

For a moment I was desperate, like perhaps a *Cataglyphis* realizing there is no entrance at the end of her long walk home. Think, think. Here is my *Cataglyphis*-like solution. Pick a large and distinct trunk, visible as far as possible from all angles. Walk around it, memorizing its lower surface. Then continue circling while widening the

A photo taken in Lake Hart, South Australia, the habitat of *Melophorus oblongiceps*, the Australian equivalent of the North African salt-pan ant *Cataglyphis fortis*. (From Rüdiger Wehner, The Desert Navigator, *2020, photograph by Rüdiger Wehner.*)

A platform from which investigations on *Cataglyphis* navigation are filmed by a Green Umbrella Team supported by the BBC during the orientation studies near Mahrès, Tunisia. *(From Rüdiger Wehner,* The Desert Navigator, *2020, photograph by Rüdiger Wehner.)*

diameter with each turn. In short, walk a logarithmic spiral, until you inevitably cross the trail. It worked for me; it will for you.

There are myrmecologists who have discovered in decades of research that a few other ant species have orientation methods at least as clever as *Cataglyphis*'s expanding loops, never mind Wilson's logarithmic spiral.

My vote for the most admirable of such devices is canopy mapping, discovered by Bert Hölldobler. While studying ants in a Kenyan rain forest, he noticed a strange behavior in foraging workers of the large predatory ant *Paltothyreus tarsatus*. As a solitary ant wandered along the forest floor during the day, likely in search of food, it would occasionally stop and align its head upward, as though examining the sky—or the forest canopy. Was the ant finding something in the sky or the canopy, or both, by which to orient its travel? Hölldobler obtained an answer with one of the most extraordinary experiments on ants ever performed in the field and laboratory. First, he took photographs of the rain forest canopy, as seen vertically from the ground. Then he captured a *Paltothyreus* colony and transported it to his laboratory at Harvard University. He established the ants in an artificial nest attached to a foraging arena into which the ants could search for food. Over the top of the arena he attached a roof that could be flooded with light from above. The roof was then covered with photos of the canopy taken in the Kenyan forest.

The *Paltothyreus* were next allowed to forage into the altered arena, find food there, and carry it back to the artificial home nest. Were they using the photographs to find their way? If so, Hölldobler reasoned, he could perceive the direction they were walking by simply turning the roof on its axis, and seeing whether the ants changed their direction in the same way and amount. He did, and they did. The *Paltothyreus* ants, in short, can both learn and use maps.

I suspect that accurate homecoming is even more important to the survival of ant colonies than researchers on the subject have surmised, and that most kinds of ants are more sophisticated in achieving homecoming than we have guessed. Consider two of the activities in which ants are insect geniuses. They can learn even subtle differences in odors and complex patterns on maps in one trial, and then remember them for a long time, even in some instances for the rest of their lives.

It has not escaped the attention of myrmecologists in years of study that foraging ant workers experience minute-by-minute a shifting earthscape of odors as they forage away from their nests. Rüdiger Wehner and his fellow researchers have established the richness of visual thought experienced by the *Cataglyphis* desert ants. It is logically possible that a similar acuity of awareness exists in these and other ants to the smells beneath their tarsi (feet).

To this end, Bert Hölldobler and I discovered a

pheromone added to the soil by leafcutter ants that helps them locate the nest entrance. We were testing the theory of another investigator that the colony odor of ants originates from glands at the rear tip of the gaster, the rearmost segment of the worker. We found evidence of the production of one or more pheromones, but the substance was an attractant that draws workers to the nest entrance, not a fragment of the colony odor.

15

ADVENTURES IN MYRMECOLOGY

ONE OF THE joys of scientific natural history is the existence of known but still largely unstudied species. Among the more than fifteen thousand ant species in the world are a few with unique features in anatomy. For researchers, both senior entomologists and students, these pose burning questions. Why are they built like that? What do they accomplish? And what is their role in the ecosystem they inhabit? As I write, I still conduct field trips, and encourage scientists to do so. We always aim to solve one such mystery or another.

For example, among the most abundant ants in the world are the dacetines, mostly very small ants with very long mandibles that snap together like spring traps. They swarm in the leaf litter and soil of temperate forests as much as they do in tropical rain forests. While an undergraduate at the University of Alabama, I set out to learn the nature of their prey. Why do they use spring traps instead of simpler, old-fashioned ant mandibles? The answer, I soon learned, is that the principal prey of the dacetine ants are collembolans, also called springtails, able to leap high into the air and away from predators upon the slightest disturbance. There exists in South American rain forests a relatively giant dacetine bearing the titular generic name *Daceton armigerum*, with anatomical traits suggesting that it is close to its small relatives. The first time I got the chance to visit *Daceton* country, in Suriname, I searched for a colony to study, found one in the forest, and got my answer concerning this Godzilla of the dacetines. The colonies nest in cavities of dead, rotting tree branches. The workers capture a variety of larger insect prey with conventional jaws and stings. Trapjaws and worldwide miniature hunting came later, in other evolving dacetines sometime in the Mesozoic Era.

On different expeditions, I have had both successes and failures. The most satisfying success was the rediscovery of the last surviving aneuretine ants. These distinctive insects, different enough for them to constitute an entire taxonomic subfamily, the Aneuretinae, were long thought to be extinct, after a strong representation among Mesozoic fossils. Then, in the late nineteenth century, two specimens of a live species, given the name of *Aneuretus simoni*, were collected in the Peridenya Garden of Sri Lanka, a nation known at that time as Ceylon.

For a young field biologist, the Sri Lanka *Aneuretus* was a prime target. Its separate subfamily rank in the taxonomic table suggested that it had a distinctive natural history. Everything that could be discovered about it would be valuable. Furthermore, its extreme rarity in collections suggested that scientists needed to hurry in its biological study and conservation.

In 1955, I journeyed to Sri Lanka and thoroughly searched through the Peridenya Garden. I found no *Aneuretus*. Same result in the forested parts of the nearby King's Garden. I then traveled to Ratnapura, in the southern part of the island. There I found a live colony of *Aneuretus*, and a dense population of the species in the rain forest around nearby Gilimale.

Another success story among the mysteries of ant

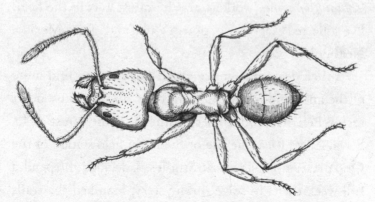

A worker of *Aneuretus simoni*, from the rain forest at Gilimale, Sri Lanka. The taxonomic group to which this ant species belongs, the subfamily Aneuretinae, flourished during late Mesozoic times, but was though to be extinct until the author found this population in Sri Lanka. (*Drawing by Kristen Orr.*)

natural history involves the strangest of all genera, *Thaumatomyrmex*, which means "miracle ant." The middle and hind part of the worker's medium-large body is normal for ants, but its head is something from a hexapod horror show. Each mandible is shaped like a pitchfork. Each consists of a flat base terminating in a row of long spikes. When the mandibles close, the spikes on one overlap with those of the other. They then curve around the side of the head, with the tips crossing at the back of the head.

What on Earth does this bizarre creature catch and eat with its formidable armament? (This is a solemn scientific question, not just an exclamation.) I decided to find out. The best way, I knew, would be to locate *Thaumatomyrmex* foragers returning to their nest carrying prey. But *Thaumatomyrmex* workers are the rarest ants in the field. I've collected exactly two, one in Cuba and one in Mexico. Neither had prey.

I then decided to make an all-out effort to find more of the ants, and follow them to their nest. Because other researchers had collected specimens in the forest at La Selva, Costa Rica, the site of the main field station of the Organization for Tropical Studies, I decided to spend a full week there, to solve the mystery. I walked the trails and cut through varieties of vegetation, focused on my solitary goal. I saw scores of other species of ants, but not one *Thaumatomyrmex*. Then, back home in desperation, I wrote an appeal in *Notes from Underground*, asking my fellow specialists working in the American tropics to please watch for *Thaumatomyrmex* in a joint effort to discover the significance of the wraparound pitchfork mandibles.

The article worked. Two Brazilians, Jorge L. M. Dinitz and Carlos Roberto F. Brandão, found one prey-laden *Thaumatomyrmex* and tracked it back to its nest. Christian Rabeling, later to join me at Harvard as a postdoctoral fellow, found a worker of *Thaumatomyrmex* carrying prey and photographed it.

In both cases, the worker was carrying a polyxenid millipede. And the mystery was solved. Polyxenids are the porcupines among millipedes; their cylindrical soft bodies are covered with dense bristles able to ward off most predators. The *Thaumatomyrmex* huntresses, by driving their pitchfork tine teeth through the bristles, are able to impale the polyxenids. When the prey are delivered to the colony, the ants display another adaptation, which we all hadn't noticed. They have rough pads on their forelegs that they use to scrub off the bristles of the millipedes. The colony then feeds on the soft bodies remaining.

Do comparable prime targets of the unknown exist that deserve special attention? There are many. I will cite three more of my favorites. First is *Santschiella kohli*, an extremely rare—at least elusive—species, known from several workers collected in the Central African rain forest of Gabon and the Democratic Republic of the Congo. The workers have enormous eyes, proportionately the largest of any ant, that make up almost half of the dorsal surface of the head, expanded enough to push the connection of the antennae to the front of the eyes instead of the rear, as in all other known ants. It is logical to assume that *Santschiella* is arboreal, living somewhere in the upper canopies of trees or shrubs, but that supposition is far from certain. Another species with very large eyes approaching the *Santschiella* condition is *Gigantiops macrops*, a native of South American rain forests. It nests in cavities in dead limbs on the ground.

So what is the meaning of the grossly oversized eyes of *Santschiella*? Better evasion of predators, or pursuit of prey during daytime travel? Or perhaps a still undiscovered form of visual communication? Yet again, better vision may enable orientation across long distances, as in many desert ants.

Another prime target of field myrmecology is the army ant *Cheliomyrmex*, a relatively rare ant of the New World tropics. Anatomically it is the most primitive of the army ants (those placed in the subfamily Ecitoninae). It is also the least studied. When colonies are finally found, their subterranean habits will make them exceptionally difficult to follow in their forest habitat. But any success in the effort to do so will be scientifically well rewarded. Ecitonine army ants as a whole are not only unique in their complex life cycles. They are among the most important predators, especially of insects and other arthropods, in the environments they inhabit.

16

THE FASTEST ANTS IN THE
WORLD, AND THE SLOWEST

EACH ANT SPECIES of the thousands on Earth has a characteristic trait called the tempo, essentially the speed at which the worker castes conduct their lives. Some species hunt for food, build a nest, and attend the queen mother in a virtual frenzy, while other species barely creep. The differences among them are enormous. Even so, both get the job done.

Fast, slow, or in between, the tempo fits the niche in which the species lives. To illustrate a high tempo and its Darwinian significance, I recommend an experiment that can be conducted on most beaches in the West Indies. Place a dollop of food, either sweet or oily, at a spot conveniently chosen (I favor the base of a palm tree), and relax nearby. After about an hour, check the bait. It will likely be swarming with ants, workers of the beach-loving *Pheidole jelskii*. Fast-moving workers arrive, feed on the bait for a while, then join a column of sister workers, some arriving and others leaving in a well-fed run back to the nest, as far as ten meters or more away. The ants will all seem frantic, as though this discovery prevents the end of their world.

Now move to another part of the beach and look for solitary workers of the same species. They are likely scouts. Any that finds a fragment of food too large to carry home will first run in more or less of a straight line, laying an odor trail in transit to call her nestmates to action. Many rush to the newly discovered food. The colony that discovers the prize possesses it.

Pheidole jelskii is a high-tempo species. It abounds on beaches, agricultural fields, airstrips, and other relatively bare environments. Its ubiquity in such habitats, maintained in the midst of intense competition, is evidence that speed counts in some habitats; high tempo can win for some species.

The fastest ants on Earth, with the highest tempo, may well be the workers and strange worker-like queens of the genus *Ocymyrmex* ("swift ant"). Its thirty-four known species have ranges that together cover most of eastern and southern Africa. They favor the hottest, most open environments, where they feed at least in part on bodies of insects and other arthropods killed by the intense heat.

Ocymyrmex have a racer's build. Their bodies are streamlined, with very long legs driven by thick segments (the coxae) at their base. The mandibles are narrow; when folded they fit tightly against the head. The spiracles of the body, the holes through which air is exchanged, are exceptionally large.

Where *Cataglyphis* ants, which travel swiftly over long distances of the desert, are the swift marathoners of the ant world, the short-range, even swifter *Ocymyrmex* are the sprinters.

During a visit to Mozambique in 2015, I had my first close look at a colony of this most extremely thermophilic and high-tempo species of ants. With a small group, I had traveled by helicopter from Gorongosa National Park to the Zambezi River delta, where I studied and collected ants in the coastal red mangrove forest. On the way back to Gorongosa, we put down at the station of an inland forest maintained as a farm of rare woods for the production of prime furniture.

It was close to midday, the air heated and literally breathtaking, as I walked out onto a baked mudflat that may have been the hottest place on the farm. There I found a nest of *Ocymyrmex*. It consisted of a single entrance opening to an underground system of chambers and galleries. Workers were running in and out of the entrance, a few thence for indeterminate distances. Some were enlarging the nest interior.

I decided to collect a few specimens from this colony for deposit in the collection at Harvard University. These ants, to this day, proved to be the most difficult to capture of any species I have ever collected at any place in the world. The ground around the nest, first of all, felt oven hot to my fingers, making concentration on fine details difficult. The ants were moving like a sizzle of water droplets in a frying pan, difficult even for the eye to keep track.

Nevertheless I insisted on collecting *Ocymyrmex*. With a vial of alcohol to receive specimens in my left hand and my favorite finely tipped Dumont No. 5 forceps in my right hand, I tried to snatch workers that for a few seconds at least slowed slightly. I am well practiced at plucking ants out of a crowd, but, challenged by *Ocymyrmex*, I caught not a single one. If I tried too vigorously, the targeted ant shot out of reach and down the nest entrance. I then moistened the outer stems of the

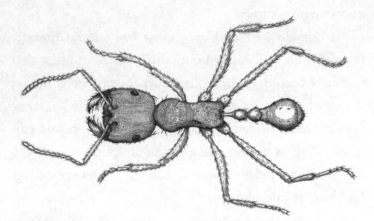

A worker of *Ocymyrmex nitidulus*, in the group of species considered the fastest running ants relative to size. From near the Zambezi River, Mozambique. (*Drawing by Kristen Orr.*)

forceps and swept them at groups of running ants in an attempt to stick one or two to the liquid. For most ants, that works. But not *Ocymyrmex*.

By this time, the heat had become unbearable for me, if not so for the ants. But I was determined to carry specimens back to Harvard. So, in desperation, I waited until several *Ocymyrmex* were rocketing around in approximately the same space, and slapped my open hand down on them. Better to have mangled ants for specimens, I reasoned, than nothing at all. This worked, albeit poorly.

Now, in order to encompass the full range of tempo in the ant world, it is necessary to leave Africa and travel to Central and South America, home of the basicerotines, the slowest and (as it turned out, a linked trait) the dirtiest known ants in the world. The basicerotines have been among the least studied and poorly understood ants in the world. Found in tropical forests, they are often called "cryptic," which means, in the vocabulary of ecology, reclusive, hard for predator and scientist alike to find.

In 1984, my close friend and collaborator Bert Hölldobler and I visited Costa Rica and the OTS Field Station at La Selva. As one of our goals we studied the natural history of *Basiceros manni*, a relatively common, albeit elusive, member of the forest ant fauna. We were also able to bring colonies back to Harvard for further research.

In almost every respect, the *Basiceros* are the opposite

of the jittery, high-speed *Ocymyrmex*. Medium in size, the workers depend for their survival on camouflage, not speed. Their opaque brown color closely matches that of the fallen leaves and mold in which they live. Undisturbed, they walk slowly. When exposed by the turn of a leaf or a fallen branch, they freeze and may remain still for minutes at a time. So difficult are the *Basiceros* workers to see that when the colony was uncovered amid leaves and soil we depended on the white color of the eggs and larvae to find their nest.

The *Basiceros* workers, unlike most ants, are ambush predators. They don't chase after prey. They stalk it, slowly, or else wait for prey to wander close enough for the ant to lunge, strike, and seize it. *Basiceros* are masters of discretion. Their tempo may be as slow as an ant species can employ and still survive.

Hölldobler and I soon learned that the integument does not merely resemble dirt. It *is* dirt. The bodies of the *Basiceros* are covered by coiled and feather-shaped hairs that efficiently collect dust and other fine detritus. They become walking dustbins.

This unique technique of camouflage led to a bizarre result at Harvard University. When we returned with live colonies, we fed the *Basiceros* wingless fruit flies and housed them in nests made of white plaster of Paris. Within several weeks, the ants had shed most of their soil-and-humus garments and replaced them with fine particles of plaster of Paris. We had unintentionally created the first pure white *Basiceros*.

In a sense, the ants had defeated us. True to their instinctive behavior and slow tempo, they had again become almost invisible, this time in the laboratory.

17

SOCIAL PARASITES ARE COLONY ENGINEERS

DOWN 150 MILLION years of evolution, predators and parasites have exploited ant colonies by a seeming infinitude of invasion techniques and con games. Almost any device that can be imagined within the scope of arthropod biology has been used to victimize some poor ant colony or another.

A colony is formed by massed bodies of cooperating ants, the nest they build, and the food they collect. It is a shaky little ecosystem. It seems almost made to order for parasites. Their rule for success is simple. Be an engineer; penetrate the relatively simple organization of the colony powerhouse and alter it to your advantage. Ants are easily fooled. They are, after all, only insects, which live by easily exploited instinct.

One such example of instinct engineering is the theft of food from the European ant *Lasius fuliginosus* by the little nitidulid beetle *Amphotis marginata*. While out foraging, the ants often distribute food they ingest by regurgitating it to one another. To stimulate the exchange, one ant uses her feet to tap another ant on her labrum, a plate covering the mouth that serves as the equivalent of a human upper lip. The ant receiving the signal from a sister responds by regurgitating a drop of liquid, which the sister drinks. In the laboratory, Bert Hölldobler has been able to get the same response by tapping the labrum with a human hair. Presumably any such light touch will suffice.

The *Amphotis* beetle, standing by the passing column of *Lasius*, make their living by expert labrum tapping. A worker ant thus stimulated is usually fooled into disgorging a drop of liquid and holding it between its open mandibles while the beetle drinks.

Scamming ants, which are among the most alert and aggressive of all insects, is easy but dangerous. It is made much safer when the parasites acquire a body odor that is neutral or even attractive to the host ant colony. Then the ants either ignore them or, perversely, carry them home as they would a sister in need of help.

Once ensconced, the mock-citizen invaders can wreak whatever instinctive havoc is needed for their own survival and reproduction. Depending on their species, they variously feed on supplies brought in by their hosts, or travel with forager ants to find food on their own. They may live on liquid food solicited from their hosts, or lick nutritious oily secretions from their bodies. Or they may scavenge the bodies of ants that die in the nest. Or, finally, in an arthropod horror story, they feed on the immature members of the colony.

The machinations of the best-adapted parasites are executed with ingenuity and precision awesome even by the standards of human engineers. Some of the most extreme examples are to be found among the guests of army ants, whose immense colonies of voracious carnivores, multilayered bivouacs, and piled-up refuse dumps find their counterparts in New York and other great cities. Carl Rettenmeyer of the University of Connecticut, working with Theodore Schneirla in pioneering studies of army ant life cycles, was the first to explore the urbanology of this insect world in detail.

What Rettenmeyer discovered, as I noted earlier, was a host of ingenious adaptations for life with ants. Some are extreme to the point of bizarre. There are silverfish, members of the insect order Thysanura, which, like the tiny limulodid beetles I had studied at the beginning of my career, ride on the bodies of ants and feed on their oily secretions. There are tiny histerid beetles (*Euxenister*) that ride on adult ants not to lick their bodies but to jump off occasionally and eat ant larvae.

Even stranger are the lifestyles of mites, tiny relatives of ticks at the outer fringes of evolutionary specialization. One, as also noted earlier, attaches itself to the inner surface of the saber-toothed mandibles of the soldier caste. Another, purse-shaped, uses its body form to clasp the base of a soldier's antenna. A third, which takes my vote for sheer strangeness, is *Macrocheles rettenmeyeri*, named for its discoverer, which grasps and sucks blood from the end of a rear foot, and thereby reduces its drag on the soldier by serving as a substitute segment of the foot it disables. This species of army ant (*Eciton dulcius*), it might be said, is the benefactor of artificial limbs.

And still, beyond the reach of anatomical inventions for the exploitation of ants, lies an even greater armamentarium of techniques to change behavior, and especially social behavior, of the victims in order to gain control of their bodies. The idea, if I might use that word, is to twist the instinctive behavior of the host enough so that the parasite, not the victim, receives the benefit of better survival and more reproduction. The trick, with ants as the victims, has been turned by a variety of cordyceps fungi, helminths (parasitic worms, nematodes), and other, socially parasitic ants.

The commonest transformation is elevation seeking, or "summiting," such that when the parasite offspring emerge from the body of the host, they can travel further and infect more hosts. The nightmarish process for *Formica* ants inhabited by cordyceps fungi is recorded by Charissa de Bekker and her fellow researchers in the following review of the social manipulation phenomenon.

The cordyceps parasites precisely time when they make their hosts leave the nest, climb the vegetation, and latch on until death. They seem to time this to the late afternoon or evening with an exquisitely precise synchronization of attachment . . . Detailed field observations further describe how infected *Formica pratensis* and *Formica rufa* move in an uncoordinated manner. Parasitized ants continuously open and close their mandibles while summiting, before attaching themselves with the head upward. Manipulated individuals also move up and down the leaf and never return to the ground. The fungus grows rhizoids that firmly attach the ant to the substrate. Subsequently, hyphal structures with infective spores emerge from intersegmental parts of the mesosoma and gaster.

Further research has disclosed that the fungus does not alter the structure of the brain. It appears instead to secrete a substance that activates and perverts programs already present, while blocking other, normal responses.

18

THE MATABELE, WARRIOR
ANTS OF AFRICA

IN THE SAVANNAS and dry deciduous woodlands of East Africa, the dominant animals are not elephants, lions, chimpanzees, and other large mammals. They are the mound-building termites. Each colony is founded by a single mother queen and her consort male. In full maturity, each mound nest contains up to two million offspring. The maximum is made possible by the prodigious fertility of the thumb-sized queen. The record, obtained with a laboratory colony, was 86,400 eggs laid in one day. Because the queen lives on average about ten years, and remains well-protected and generously fed in the mound nest, she may produce something like 100 million offspring in her lifetime.

The food of the colony is a symbiotic fungus grown in sponge-like masses in special garden chambers. The air of the nest interior is freshened by tunnels built throughout the nest, with the old air driven out by heat from the bodies of the termite inhabitants. The substrate of the gardens is built from chewed fragments of dead vegetable material collected from the ground around the nest.

As the colonies grow from the equivalent of villages to cities, their nests increase in size from refrigerators up to city buses, as much as three meters high and ten meters across. From most of the largest nests grow a variety of plants that includes shrubs and even trees. Birds alight to feed and build nests, and some smaller mammals climb up to the top to scan the surrounding terrain.

If you go out and sit next to a mound on a quiet night, you will hear a continuous soft hiss, raised by the patter of feet of thousands of workers out of the nest, all large males as it turns out, searching for dead plant fragments to feed the symbiotic fungi.

Massed inside the nest with its fungus crop, each colony is a prime target for predators. Eons of evolution have, in response, produced several layers of defense by the termites. The outer crust of the nest is a strong shell of soil chemically treated by its inhabitants. If nevertheless penetrated, the colony launches three emergency responses. The exposed workers rush deeper into the nest interior and out of sight. Soldiers rush outward to guard the breach. They are built for combat, with heads hard and sclerotized, like helmets, and long, needle-tipped mandibles projecting forward and closed with powerful jaws.

The colony is well-organized in an emergency, for example by the rupture of the nest by a pangolin or scientific investigator. When the danger has passed, the soldiers pull back and workers crowd in to repair the damage, the third emergency response.

If you take an adze or shovel and dig into the side of a small nest densely populated near the surface, and especially if you continue digging until you come to one of the gardens, you will first see swarms of pale white workers. Instantly they begin to disappear into the dark recesses of the nest interior. Come back an hour or two later, and a large force of workers will have returned, to cover the breach with moistened pellets of frass. The soldiers have mostly left. Come back the next day, and the damage will be fully repaired.

A virtual law of scientific natural history is that for each potentially rich prey there evolves, given enough time, a predator specialized to exploit it. (This rule includes parasites, which may be defined as predators that eat their prey in units of less than one.) One specialist of the mound-building termites I have been able to study in Gorongosa National Park in Mozambique is the matabele ant, with the scientific name *Megaponera analis* (see the frontispiece portrait). Unusually large in size for an ant, encased in a heavy chitinous armor, and able to travel fast in organized groups, it seems programmed to feast upon mound-building termites. The local common name is appropriate: it refers to the fierce "Men of the Long Shield" of western Zimbabwe.

Matabele ants have a frightening sting, the worst I have ever encountered in ants, evidently designed to send a lifelong message to any bird or mammal inclined to eat one. When at Gorongosa I picked a worker up (and immediately thereafter vowed it was the last one) to examine it closely, the matabele first gnashed its jaws impressively, then twisted its gaster (the part of its body behind its waist) forward and thrust a long sting into the flesh of my index finger. On a pain scale I would rank it close to a hornet, perhaps two or three hornets. I dropped it unharmed, one of the few ants in my long entomological career to defeat me single-handedly.

Is the matabele sting the worst of any in the ant world? The most serious competitor is the giant *Paraponera clavata* of Central and South American rain forests. An expression of its power is that in part of its range its common Spanish name is Dos Semanas, meaning that it takes two weeks to recover from the sting. At least one indigenous tribe in South America has used *Paraponera* in its manhood ceremony. The great myrmecologist William Morton Wheeler is rumored to have fainted after being stung by one on Barro Colorado Island in Panama. Fortunately, colonies are small, and workers are neither aggressive nor fast-moving.

A matabele raid on a termite nest is one of the most

dramatic wildlife spectacles of Africa. Even to witness a column on the march to collect termites is worth a tourist's excursion away from camp. Its conclusion, the end of the battle, reveals one of the most astonishing phenomena known to me in tropical biology. The goal of the raiders is not to steal the fungus growers' gardens. They are after the bodies of the gardening termites themselves, and in particular those killed in battle.

For the matabele, war is a hunt for dinner.

A matabele raid starts with a single scout that finds and inspects a termite mound within marching distance of its own nest. She evidently is searching for some entrance or other, or an accidental crack by which to gain access to the termites. If successful, the scout runs back to her nest, in a straight line, laying a chemical trail from her body along the way. The substance is a powerful pheromone that draws out a battalion of huntresses. They form up and run in a column several ants wide all the way to the break in the termite mound.

The raiders are like the *impi*, the human Matabele fighting regiment. There is nothing of the dallying and running around and occasionally turning back toward the home nest that characterizes most kinds of ants. There is in the matabele ant an all-or-nothing commitment. Unity is a necessity. Almost at once, upon the break-in, the matabele are met by an equally fierce swarm of termites. The massed raiders are unfazed, and simply overrun the defenders. They kill and gather up the dead termite soldiers, along with a few ordinary worker termites they encounter, and then march back home. A single matabele ant worker can carry as many as ten of the dead in its mandibles.

The matabele, so far as I know, do not occupy the mound after their victory. The raid is everything.

19

WAR AND SLAVERY
AMONG THE ANTS

IN 1854, IN his classic *Walden, or Life in the Woods*, Henry David Thoreau described what he believed to be a war between two species of ants.

> One day when I went out to my wood-pile, or rather my pile of stumps, I observed two large ants, the one red, the other much larger, nearly half an inch long, and black, fiercely contending with one another. Having once got hold they never let go, but struggled and wrestled and rolled on the chips incessantly. Looking farther, I was surprised to find that the chips were covered with such combatants, that it was not a *duellum*, but a *bellum*, a war between two races of ants, the red always pitted against the black, and frequently two red ones to one black.

Was it war that Thoreau observed, as he reasonably thought, or was it something else? It has been my own experience over decades with ants in North America that battles of this kind between two very different species are often slave raids—in this case, a red slave-maker, most likely *Polyergus lucidus* or a member of the *Formica subintegra* group of species, in a raid against a vulnerable black species, possibly the common *Formica subsericea*.

Raids and resistance in fierce combat occurs, but slavery in most cases is not at all the same as slavery in human beings. It is more like the capture and domestication of wild animals.

The workers of ant species specialized to be slave-makers are programmed by instinct to raid colonies of otherwise similar species. They have a single target: the pupae of the colony being attacked. These captives, usually after a bitter fight by the adult residents, are carried unharmed back to the raiders' nest and allowed, following a few days or weeks, to eclose from the pupal case as fully formed adult ants. Ants everywhere, so far as we know, have a trait that allows them to be turned into slaves: newly emerged adults acquire the odor of the colony in which they eclose. The result is that they accept the raider workers as sisters, and the raiders accept them in the same way. To add devoted adults to the worker force by whatever means conveys a major advantage to the colony when in competition with other colonies of the same species.

The method by which colony odor is acquired by ants in general was discovered by the pioneer behavioral biologist Adele M. Fielde around the beginning of the twentieth century. By this means, it became possible to create colonies whose members vary radically in size and anatomy. Large spiny species, for example, can be combined with small, smooth-bodied ones.

In the north temperate zones of North America, Europe, and Asia, ant slavery is common, especially in the subfamily Formicinae. One group of raider species, those comprising the genus *Polyergus*, bright reddish brown in color, are strongly developed in behavior and anatomy for their trade. They wield their saber-toothed mandibles with speed and violence during the raids, which are regularly conducted against dark-colored *Formica*.

Several combinations of raider versus raided within the rich ant fauna of New England are possible candidates for a raid that might have been what Thoreau called war. But we may never be sure, especially since he described one species as much larger than the other. For all his powers as an observer, he did not collect specimens of the ants for future entomologists to study and identify. This omission is unfortunate, for among his circle of friends, in addition to Ralph Waldo Emerson, was Louis Agassiz, who at the time of Thoreau's ant *bellum* was building and stocking Harvard's Museum of Comparative Zoology, today the custodian of the largest scientific collection of ants in the world.

Meanwhile, slavery among ants has been documented

in the field by me and others to include an astonishing variety of parasitic or warlike behavior. In Yosemite National Park, while a graduate student at Harvard, I discovered a colony of the slave-maker *Formica wheeleri* with two species of slaves. One was engaged in a raid when I came upon it. Running with the raiders were workers of a third species of *Formica* which evidently were serving as janissaries—assistant soldiers—of the marauders. When I dug into the nest, I found yet a fourth species of *Formica* huddled with the eggs, larvae, and pupae of the marauder and evidently serving as nurses.

Over thirty years later, while giving an address to the assembled superintendents of the U.S. National Parks, I "confessed" that I had excavated an ant nest in Yosemite, and asked forgiveness for what I now recognized as a transgression. In introducing me for a second address I gave several years after this event, the Director of the National Park System issued forgiveness and gave me a nicely decorated permit to collect *one additional* worker of the same marauder species in Yosemite National Park.

Is slave-making a dead end in evolution for the species that adopts it? Not quite—but it can degenerate still further. A striking example is found in the socially parasitic genus *Strongylognathus*, which occurs across Europe and Asia. Most *Strongylognathus* species conduct standard slave raids, using saber-toothed mandibles to subdue their victims. In one species, *Strongylognathus testaceus*, the workers have lost their warrior spirit. Instead, the newly mated *Strongylognathus testaceus* queen simply moves into the host colony and sits near the host queen. Thereafter, the host workers take care of the parasite as well as they do their own mother. Daughters born of the parasitic queen are amicable toward their hosts but undertake no work.

Yet another step into the world of slave-making has been taken by the American slavemaker ant *Formica subintegra*: it is an expert at propaganda. The Dufour's gland emits a pheromone that alerts nestmates to the presence of danger. In *Formica subintegra*, the gland is enormous, filling as much as a third of the volume of the gaster, the large hind segment of the body. The workers spray their secretions at resident ants during slave raids. The secretions are so strong that they create panic in the colony being attacked, making it easier for the raiders to penetrate the brood chambers and carry off pupae for future slavery.

20

THE WALKING DEAD

EVERY CORPSE IS an ecosystem. Each fallen bird, landed fish, beached whale, decomposing log, plucked flower is destined to change from a conglomerate of giant molecules, the most complex system in the universe known, into clouds and drifts of much smaller organic molecules. The process of decay is driven by scavengers, in nature beginning with vultures and blowflies and ending with fungi and bacteria.

What do ants do with their dead? If a colony member is badly injured in the field, it is in many species carried home and eaten. If injured only moderately, it may be allowed to live and heal. Most ant warriors that die in battle outside the nest never return. They instead fill the jaws and beaks of predators.

An ant that dies from old age or disease inside the nest simply comes to a standstill or else falls to the side with her legs crumpled up. In most cases, she is allowed to stay in place. After at most a few days, a nestmate picks her up and carries her out of the nest or to a refuse pile in one of the chambers within the nest. In this cemetery chamber is also dumped miscellaneous refuse, including the inedible remains of prey. There is no ceremony.

It occurred to me early in my studies of chemical communication in ants that the bodies of dead are likely recognized by the odor of their decomposition. Of all the substances uniquely present in dead insects, one or more must be the signal that triggers corpse disposal by ants. If live ants demonstrably use such molecules to release other instinctive social behavior in the service of the colony, why not in death also?

It was my good fortune at the time to find a published account, understandably obscure, that identified substances found in dead cockroaches. Using this work as a guide, I set out to learn what chemicals stimulate necrophoric (corpse removal) behavior in ants.

As a first step, I made extracts of decomposing ants. I put droplets of this material on "dummies" of dead ants made of flecks of balsam about the size of workers. When these were dropped into nests of laboratory colonies of harvesting ants, each was picked up and taken speedily to the refuse pile. So now I had a working bioassay, the essential step in biological experimentation. At the same time, I acquired synthetic, chemically pure samples of the decomposed cockroaches. For a while the laboratory smelled faintly of a mixture of charnel house and sewer. (Two of the substances, for example, were the terpenoids indole and skatole, elements of mammalian feces.) Most of the substances tested caused excitement and aggressive circling by the ants, but did not result in immediate removal. Where bits of balsam treated with odorous substances were attacked or simply ignored, those carrying indole or skatole were picked up and carried to the cemetery.

There is no procedure more pleasing to a biologist than an experiment that works. This one was successful, at least for Florida harvester ants, and I repeated it for visitors to witness until I grew bored. So I asked a new question: What would happen if I daubed a live, healthy worker with one of the funereal substances?

The result was gratifying. Worker ants that met their daubed nestmates picked them up, carried them alive and kicking to the cemetery, dropped them there, and left. The behavior of the undertaker was relatively calm, even casual. The dead belong with the dead.

The daubed ants did what you and I would do if turned into zombies: We would take a bath. It should be no surprise that this solution is also used by ants that suffer an unwanted material on their bodies. They pull the flexible outer segments of their antennae, the funiculae, through comblike structures on their forelegs. They lick as much as possible of their body and legs with their pad-shaped tongues. They curl the gaster, the rearmost part of the body as far forward as possible and wipe and wash it. They take a typically ant bath.

Then they return to the main living quarters of the nest. If enough of the necrophoric substance on their bodies has been removed, they are accepted back into the nest. If not, they are returned to the cemetery by their nestmates. They continue to clean themselves, and perhaps others assist them. They wait. In time, if the contaminants are removed or sufficiently dissipated, they rejoin the living in full.

21

TINY CATTLE RANCHERS
OF AFRICA

THE PHYSICALLY STRANGEST ants in the world may well be the five known species of the genus *Melissotarsus*, distributed variously across parts of tropical Africa, Madagascar, and the Comoros Islands. Absurdly squat in physique for ants, with a solidly welded midsection of the body covered by finely wrought longitudinal grooves but none of the sutures or seams typical of ants, plus massive coxae attaching the legs to the body, oversized heads, and short seven-segmented antennae (where most kinds of ants have ten to twelve), a *Melissotarsus* queen or worker is recognizable at a glance.

I have never seen a live *Melissotarsus*. I expect to remedy that shortcoming when I next visit Gorongosa National Park in Mozambique. I will be accompanied by Piotr Naskrecki, the park biologist and by record of knowledge and accomplishments in the field one of the several best naturalists I have ever known. He promises me a *Melissotarsus* colony.

We both know finding it may be a complicated and difficult task. *Melissotarsus* lives only in the healthy wood and under the bark of living trees, carving out the galleries and chambers that make its nest.

So far as known, the workers never leave their stronghold. Instead, they are specially adapted to travel efficiently through the tunnels they have dug. They move on their fore and hind legs, while raising the middle legs to touch the roof as they travel. In a word, they shimmy their way through their homes. It is said that when removed from the nest and placed in the open on a flat surface, with no roof to touch, they are unable to walk. We shall see.

If *Melissotarsus* colonies are largely confined to the live xylem in which they have cut their nests, what do they

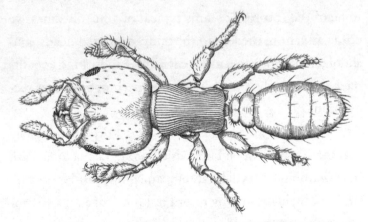

A worker of the "cattle-raising" ant, *Melissotarsus beccarii*, of tropical Africa. (*Drawing by Kristen Orr.*)

eat? The answer is perhaps the strangest thing about them. They herd "cattle."

Within the *Melissotarsus* nest live armored scale insects, protected and tended by the ants. A symbiosis of this kind is common and widespread among the ants. Scale insects—mealybugs, aphids, and other homopterous insects—in exchange yield droplets of excrement rich in sugars and amino acids. In the usual, simpler form of the exchange, worker ants travel from their nests to homopterans that are in the act of feeding on plants. In more advanced partnerships, the homopterans live as guests within the ant nests, feeding on the roots and other protected parts of plants. The guests provide a constant source of liquid food for their ant hosts. When protein is scarce, because the ant huntresses are less successful, the colonies have the option of killing and eating their homopteran guests.

The armored scale insects that live with *Melissotarsus* colonies do not produce nourishing food in their excrement. They, evidently, are killed and eaten by their hosts, prudently, thus providing protein directly. *Melissotarsus* are herders, not farmers.

22

TRAPJAWS VERSUS SPRINGTAILS

THE MANDIBLES OF dacetine ants function like mouse-traps, even though their structure is radically different. Armed at their tips with spikes and, along the inner margins, rows of teeth as sharp as surgical needles, they strike prey with one of the fastest organic movements known in the living world.

Upon sensing prey nearby, a dacetine huntress, whether a tiny *Strumigenys* creeping through soil and leaf litter or a relatively giant *Daceton* patrolling the canopy of a tree, throws open her mandibles as widely as possible. (In some species, the spread is 180 degrees or more.) The mandibles are pulled tight by muscles that fill the back of the head. They are locked in place by a catch on the labrum. When the ant pulls her labrum back, the mandibles are freed, slamming their blades together, teeth first, on whatever may lie between them.

The strike of a dacetine is too fast for the naked eye to follow. Video analysis has revealed that the nerve impulse from brain to labrum, followed by the release of mandibles from the labral catches, consumes only 5 thousandths of a second. The strike that follows is completed in 2.5 thousandths of a second.

Slow movement followed by explosive quickness is the hallmark of the hundreds of species that compose the taxonomic tribe Dacetini. The smaller dacetines are also among the most common ants throughout warmer parts of the world.

Recently, laboratory experiments have revealed that an equivalently explosive strike occurs in a species of the primitive and widespread ant genus *Odontomachus*. My first encounter with one of this species, *Odontomachus brunneus*, occurred when I was a boy excavating a colony in Alabama. I experienced the trapjaw's double weapon: the shock of the trapjaw bite followed almost instantly by the burn of the sting.

A similar trap has been evolved independently by a relatively primitive species, *Mystrium camillae*, of Madagascar. The tips of the mandibles are pressed together, then slid very tightly one across the other and released explosively—like the snapping of middle finger pressed against thumb. This movement by *Mystrium* is about 90 meters per second, or more than 200 miles per hour in conventional distance and time. When applied to the very small size of an ant, this per-meter velocity stands as another all-biology record.

Dacetines occur on the land over most of the world. They reach maximum diversity and abundance in tropical and warm-temperate forests. The northernmost record was made in 1950 by the late William L. Brown, a distinguished myrmecologist and my mentor at Harvard University. For the true ant-cognoscenti, who may wish to break the record, the colony he found was a *Strumigenys* living in a clump of accumulated grassy soil at the base of the right rhinoceros statue facing outward at the main entrance of Harvard's Biological Laboratories. The ants were gone when I arrived at Harvard in 1951, and none have returned in the nearly seventy years of my continuing tenure at Harvard.

Bill Brown, in 1950 a PhD student at Harvard, assisted by his wife, Doris, reorganized the ant collection of the university's Museum of Comparative Zoology, from the condition left at the death of William Morton Wheeler in 1937. His enthusiasm for ants and dacetines in particular was infectious for a newcomer. He soon had me working for him.

"Wilson," he wrote while I was still a student at the University of Alabama, "there are a lot of dacetine species in the southeastern United States where you live. Some are bound to be new to science. I need to see all I can get. Collect what you can in Alabama and send them to me."

Then he invited me into broader research on ants: "Wilson," he urged, "find out what dacetines eat. What do they hunt and catch creeping around with those weird mandibles?"

Even without Brown's encouragement, I would have been fascinated by these little ants of our shared delectation. They posed burning questions. What is their place in the biosphere? What do they catch and eat? But there was no way a scientist could just go into the field and follow them around, recording their diet. It was difficult enough simply to find a nest of dacetines, let alone follow the little huntresses as they forage through fallen leaves and beneath the bark of rotting logs.

I solved the problem with what I came to call the cafeteria method: instead of following ants into the environment, I brought the food to the ants.

My cafeteria is an artificial nest, a rectangular block of plaster of Paris roughly the size of a man's shoe, containing two deep chambers side by side connected by a narrow channel which allows the ants to run from one side to the other. Each chamber is topped by a glass pane. On top of one is set another pane, red in color, which the ants see as dark. A colony of ants placed in either chamber will choose the darker one and turn it into a nest, as much as circumstances allow.

Now comes the experiment. Into the lighted chamber place a sample of soil, leaf litter, decaying wood, or other micro-habitat containing a variety of potential prey, including especially those encountered in the vicinity of the site where the ant colony was found. Which ones the ants select can be seen as they carry their booty into the nest chamber. Mites, spiders, schizomids, centipedes, millipedes, nematodes, earthworms, flies, beetles in great array, termites, and other kinds of ants are part of an almost endless menu.

The cafeteria method, with the ants selecting their own meal in a simulated piece of their natural environment, has worked remarkably well in practice. I've used it to cast light on a separate mystery of the forests of tropical America that particularly fascinated me: Why are colonies of very small workers of the ant genus *Pheidole* everywhere so numerous and populous in the forests? Their prevalence was likely important to the harmony of the ecosystem, but how? What are they doing that makes them so successful? Is it the nest sites they choose, then alter? From my field experience with many colonies in Central and South America, I doubted intuitively that the nest sites mattered. The ants chose places that were already optimal for them, rather than creating new sites. Then perhaps food: that might be varied and specialized enough to have an impact on the living environment. In search of an answer, I used a cafeteria experiment.

The answer I got was a surprise. Oribatid mites! The *Pheidole* gathered these small, spherical, inoffensive fungus-eaters in large numbers, like pumpkins from a September garden. Each ant was able to carry home, singly albeit often with difficulty, a live mite, a tiny ball with moving legs, into the nest—eventually to be shared as food with her nestmates.

What I discovered about the prey of tiny dacetines was even more surprising. The dominant genus, *Strumigenys*, collected and brought to the artificial nest a wide variety of small, wingless, soft-bodied arthropods, including collembolans. They passed over mites of all kinds and collembolans of the family Poduridae, the latter well-known to carry poisonous chemical defenses against predators. The favored prey, out of a wide variety available, were collembolans of the family Entomobryidae. These small insects are famous for their ability to jump out of danger when they encounter their enemies, which in the grassroot jungles of the world are many.

I discovered that the trap-jawed *Strumigenys* and the leaping entomobryid collembolans are in constant conflict, each using its own explosive devices, one to capture, the other to avoid capture.

The species of *Strumigenys* have adapted a wide variety of hunting among them, employing their trapjaws in different methods of stalking prey. The workers of *Strumigenys louisianae*, a common species throughout the southeastern United States, are bolder and more direct than most other small species in the genus. Their confidence is permitted by their more efficient mandibles, which are long and formidably spined. The worker *Strumigenys*, approaching a collembolan, moves slowly and cautiously. It spreads its mandibles to the maximum angle and in so doing exposes two long hairs rising from the paired labral lobes. These hairs extend far forward of the ant's head and serve as tactile range-finders for the coming mandible strike. When they first touch the prey, the body of the prey is well within reach. The sudden and impulsive snap of the mandibles literally impales the collembolan on the apical teeth, so strongly that drops of hemolymph, insect blood, often well out of the punctures. If the collembolan is small relative to the *Strumigenys*, the ant lifts it into the air, and may then sting it. All but the largest collembolans are quickly immobilized by the ant's strike and sting. Their struggle is feeble and short-lived.

The worker of *Strumigenys membranifera*, with much shorter mandibles, is more circumspect than the worker of *Strumigenys louisianae* when stalking prey. As soon as the ant becomes aware of the presence of a collembolan, it "freezes" in a low, crouching posture and holds this stance briefly. If the collembolan is to its back or side, the worker then turns very slowly to face it. Once it is aligned in this way, the ant begins a forward movement so extraordinarily slow that often it can be detected only by persistent and careful observation. Several minutes may pass before the ant finally maneuvers to within less than a millimeter's distance and adopts a striking position, and it may remain in this position for as much as a minute or longer. Unlike *Strumigenys louisianae*, the *Strumigenys membranifera* open their mandibles to only a triangle of about 60 degrees. Tactile hairs are present and eventually come to touch the prey. The mandibular strike is as sudden as that of the *Strumigenys louisianae*, but since it is usually directed at an appendage it does not have the same stunning effect on the collembolan. The insect often struggles violently to escape, but the ants are very tenacious and retain a fast grip until they are able to sting the prey into immobility.

To summarize, *Strumigenys louisianae* relies on a comparatively swift approach to its prey followed by a fixed-action pattern that can be characterized as strike–lift–sting, with the last element occasionally being omitted if the prey is small. In contrast, *Strumigenys membranifera* employs a more cautious approach followed by strike–hold–sting, with the last element inevitable. The *louisianae* pattern is apparently typical for long-mandibulate dacetines generally, while that of *membranifera* is typical for the short-mandibulate groups.

The ecological significance of the difference between the two groups of dacetines is in my judgment the following: the *membranifera*, requiring less space for the operation of the mandibles, is generally associated with cryptic foraging. The Japanese myrmecologist Keiichi Masuko has discovered an extreme version of this short-mandibulate technique in *Strumigenys hexamera*. These bizarre little ants are the ultimate ambush hunters. The mandibles are directed slightly upward from the plane of the head, and the dorsal apical teeth are especially long and sharp, enabling the ant to strike with particular effectiveness at objects looming directly above its head. The *Strumigenys* forager hunts a great deal in small crevices within the soil. Because of the tightness of the passages, it usually encounters the prey in front. The ant immediately crouches and freezes, pulling the antennae completely back into the scrobes that line the sides of the head. The mandibles remain closed. Even though the target, such as a collembolan or small centipede, may be very close by, the ant never moves toward it. Instead, it remains perfectly still for a period of twenty minutes or longer, waiting for the prey to step on its head. Then, with a sudden upward lift and snap of its mandibles, it impales the victim on its long apical teeth.

Entomologists classify the collembolans favored by the smaller dacetines as members of the family Entomobryidae. They are tiny, wingless, and soft-bodied, ideal food for predators such as the small dacetines. Entomobryid collembolans occur almost everywhere in natural environments on the land as well as on the pleuston, to use the word employed by ecologists for the strange inhabited surface film of freshwater ponds and lakes. They have been found in high reaches of Mount Everest, where no ants live (small jumping spiders are their likely predators). They have also turned up in samples of water drawn from Lake Vostok in Antarctica, which is covered by a permanent thick layer of ice. Breeding populations of entomobryids live all around us. They may even be found on snow piles in North America during sunny winter days.

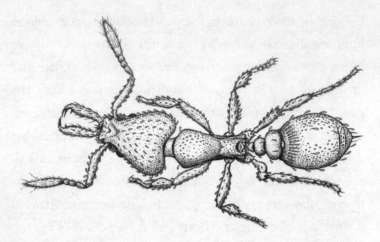

The classic example of trapjaws used by ants to capture fast-moving prey is displayed by the little *Strumigenys louisianae*, found through most of the United States. *(Drawing by Kristen Orr.)*

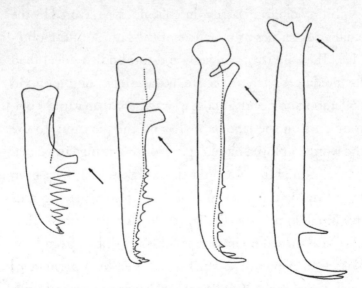

The trapjaws of ants in the tribe Dacetini have lengthened in evolution. *(Modified from W. L. Brown and E. O. Wilson, 1959.)*

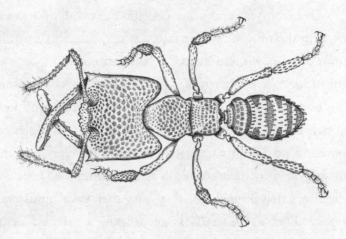

A worker of the Madagascar ant *Mystrium camillae*, which snaps its jaws like fingers pressed against thumb and violently released. (*Drawing by Kristen Orr.*)

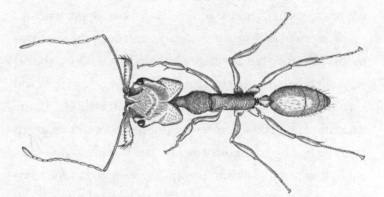

A worker of *Odontomachus brunneus*, a large predatory ant with trapjaws. (*Drawing by Kristen Orr.*)

Entomobryids and other collembolans could not coexist with such ingenious predators as dacetine ants for millions of years without evolving countermeasures. Some of the major collembolan groups have evolved noxious, poisonous body substances sufficient to keep collembolan-hunters at bay. Entomobryids and a few other collembolan lineages have invented another, entirely different device. They are, as suggested by their common name, springtails. Like the dacetines that hunt them, they have their own mousetrap device. Each carries a "tail," technically a furcula, a stiff device tucked firmly along the undersurface of the body and attached to it at the rear end. When a collembolan is mortally threatened, the furcula is released. It swings downward violently, flinging the body upward and outward, beyond the reach of any predator of similar size that approaches it.

I haven't tried to measure the trajectory of an entomobryid jump, but a reasonable guess would be that if the collembolan were the size of a human being, it would travel as far as a football field and half as high. If the dacetine ant strikes before the jump, and sinks its teeth into the body of the collembolan, the furcula is powerful enough to carry both into the air. I've watched in the laboratory as this event occurred. The dacetine almost always hung on, and the only problem for the ant was that it had to take a longer walk home.

23

SEARCHING FOR THE RARE

LET HIM WHO boasts the knowledge of actually existing things, Saint Basil the Great wrote, *first tell us of the nature of the ant.*

In meeting Saint Basil's challenge literally, myrmecologists have explored unusual environments throughout the world and enjoyed unique challenges and physical adventure at its very best. Students of ants have stories to tell, not just to fellow specialists but to anyone interested in the challenges of natural history.

One such story, scientifically among the most important to date, was the search in southwestern Australia for *Nothomyrmecia macrops*, the most primitive living ant species known at the time. In 1932, a young woman naturalist had collected the first two specimens while on a horseback expedition across the sandplain heath between Esperance to the west and the Nullarbor (translation, "treeless") Plain to the east. Myrmecologists were fascinated by its wasplike anatomy. Might *Nothomyrmecia* colonies possess social behavior as primitive as its anatomy, and provide us with clues to the origin of ant societies from ancestral wasps, which we estimated to have occurred over a hundred million years earlier? We needed to study living colonies in order to find out. In the summer of 1954, I went to the collection site with three companions. One was a mechanic to drive our overland vehicle, the second a naturalist from Perth, and the third was Caryl Parker Haskins, in his mid-forties a prominent geneticist, public official, and author of the best-selling book *Of Ants and Men*. Caryl was an ardent myrmecologist and, as it happened, the leading expert on the ferocious bull ants of Australia, the closest known relatives to the nothomyrmecines.

A camp at night in the boundless Australian outback, with a cooling breeze, whining dingoes circling the campfire in search of scraps of food thrown away, a sense of unknown variety of life in all directions—wilderness in the best sense of the word. I thought, it will never get better than this, especially if we could find *Nothomyrmecia*.

As night fell on our second day in the field, Caryl Haskins and I walked out into the dark, flashlights in hand, hoping that *Nothomyrmecia* workers were nocturnal and came out then. We found none, and were soon lost. The sandplain heath had no trails and no local features that offered compass direction. When it became clear that we would have to wait until dawn to find our way back to camp, Caryl looked about for a rock of the right size and shape to serve for a pillow. He carried it into a clearing, lay back, positioned his head, and fell asleep. I spent the night walking around him in widening circles, flashlight in hand, hoping the *Nothomyrmecia* were nocturnal and I might find workers foraging away from their nest.

No luck, either then or afterward. During the four-day visit, we collected a couple of ants new to science, as well as rare known species, and enjoyed the grandeur of the heathland fauna and flora. In that time we glimpsed one automobile pass along a distant road. On another day a white stallion trotted up close, out of nowhere that we could tell. It stood looking gravely at us for several minutes, then trotted away to disappear on the sand-plain heath.

We found no *Nothomyrmecia macrops*, not a single worker. We were in the right place—double-checked. How could we have failed? Later, after we returned to the States, the news of American scientists searching for what now was called the "dawn ant" spread widely in Australia. It stirred local entomologists into an effort of their own. Australia's most famous ant should be rediscovered by Australians.

Success, when it came, revealed why we had failed. Robert W. Taylor, a New Zealander working in Australia, who had earned a PhD under my direction at Harvard and was then employed as a government researcher at CSIRO in Canberra, organized an expedition to the original collection site with the intention of scouring the habitats until they came up with the dawn ant.

They left Canberra in the early winter of southern Australia, traveling by automobile from Canberra south-west to Adelaide on the southern coast. On the first night, they camped a short distance beyond the city, in a forest of mallee, a species of dwarf eucalyptus.

There was a chill in the air around the campfire, but not yet sharp enough to halt all activity by the local insects. When dinner was finished and the others were settling down, Taylor ventured into the mallee to collect specimens of the local ant fauna. A short time later, he ran back into camp shouting, "I got the bloody bastard! I got the bloody bastard!"

Nothomyrmecia macrops had been rediscovered, and the Adelaide camp area became the center of *Nothomyr-mecia* field studies thereafter. Colonies were collected for laboratory research. The species, in time, became one of the best known of living ants. The likely early stages of ant evolution were then more securely estimated.

With a relatively dense population of *Nothomyrmecia* now located, we could ask, Why did my fellow searchers and I fail to turn up even one specimen in our own efforts? The answer soon became clear: we had been searching at the first locality, east of Esperance, in warm to hot sum-mer weather. Taylor and his fellow searchers had chosen comparatively chilly, close-to-winter weather for their efforts. We now understand that *Nothomyrmecia* succeeds at least in part because it is a cold-weather specialist. It is nimble enough to find and capture insects and other arthropod prey that are more chilled and less nimble.

At the climatic opposite extreme to *Nothomyrmecia* are the *Cataglyphis* ants of Eurasia and Africa, which thrive in hot weather on burning sand, and collect prey that are killed or immobilized by the extreme heat. A close parallel to *Nothomyrmecia* in climate specialization among ants in the North American temperate region is the cold-weather ant *Prenolepis imparis*. Its colonies are very active on milder days in the winter, forming lines from the nest out into hunting groups, but in the summer they retreat into deep, cool tunnels excavated for the purpose.

As Charles Darwin discovered during his month-long visit to the Galápagos in 1835, large, distant islands of ancient age are often rich relics of endemic species—plants and animals of a kind found nowhere else. There is romance of a kind, and potentially science at its best, waiting for biological specialists to discover and for the time to study them. So it was in 1969, when William L. Brown was the first ant expert to visit the remote Mascarene archipelago, including the island of Mauritius, home of the long-extinct dodo.

Would there be ant equivalents of the dodo among the Mascarene Islands, and particularly in the surviving natural environments of Mauritius?

The prime locality to search for the last remnants of this island's endangered fauna is Le Pouce, an elongated massif with a topside plateau covered with low, gnarled native forest. After one only partially successful trip to the plateau, Brown decided on a second effort, which he later described as follows:

On the first of April, though I was scheduled to leave on an evening flight to Bombay, I tried Le Pouce again. A telephone call to Mr. J. Vinson, who had collected some material, convinced me that the main path on the Le Pouce plateau should not be avoided. I arrived there in the afternoon; the day was heavily overcast, threatening rain on the peaks, and it took me about an hour to walk up to the plateau. Whereas the sunny Sunday in the scrub shade had yielded almost no ants foraging, I now found foragers on foliage and on the hard-packed earth of the trail every few meters of the way. These were mostly *Camponotus aurosus* and *Pristomyrmex* spp. (= *Dodous*), native Mauritian species. Before long, on the trunk of a small tree by the path, I found a sparse trail of bright red ants climbing the bark. Closer examination revealed these to be predominantly the ectatommine I had collected on the previous Sunday, but interspersed with these were workers of *Pristomyrmex bispinosus* which, with their gasters partly curled under, looked remarkably like the ectatommines. It is hard to avoid the impression that some kind of mimicry involves these two species in this habitat. The ectatommines ascending the trunk nearly all carried in their mandibles whitish spherical objects that proved eventually to be arthropod eggs—probably spider eggs. I climbed the tree which was only about 5 meters high, and soon found the nest about 3 meters up. Where two of the gnarled branches crossed, a thick pad of lichens surrounded the place where they touched. Forcing the branches apart, I found a rotted-out pocket, evidently caused by their rubbing together in high winds. The cavity extended downward several centimeters into one of the branches, and it was full of the ectatommine ants with brood and many of the round white arthropod eggs; I estimate that I removed or saw at least 200 workers, and there may have been more.

Afterward, in an incident about which he did not write, Brown attempted to climb the small peak at the end of the plateau. Lightning struck 20 feet upslope, then heavy rain turned the ground into slippery mud. Brown fell and slid down the slope toward the brink of a high cliff. Hanging on to a small bush for a few minutes, he finally regained his purchase and worked his way back down to the trail and then to Port Louis.

The bright red ant turned out to be an undescribed *Proceratium* (*P. avium*), notable for its aboveground foraging for eggs, in marked contrast to the other, wholly subterranean members of the genus. Also, the enlargement of its one-faceted eyes suggests that *avium* emerged as an aboveground forager late in evolution, after the arrival of its ancestral stock on the remote island of Mauritius.

Bill Brown's Mauritius ants, secured at the risk of his life, did not happen as a consequence of his targeting species of great antiquity. He traveled to the Mascarenes as an effort of pure exploration. What species of any kind, he asked, might be found?

The planet contains many other places of which the question may be asked by new generations of scientists, "What species of any kind might be found?"

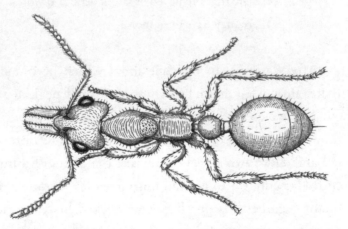

A worker of *Nothomyrmecia macrops*, the Australian winter ant, which has been judged one of the most primitive ant species in the world. (*Drawing by Kristen Orr.*)

24

AN ENDANGERED SPECIES

IN 2011, I led a team of field biologists on an expedition to Vanuatu, formerly the condominium of the New Hebrides ruled jointly by France and Great Britain. Vanuatu had become a new, independent democracy. Lloyd Davis, Kathleen Horton, Christian Rabeling, and I were able to collect specimens intensively on the large, northern island of Espiritu Santo, and the capital, central island of Efate. Fifty-seven years earlier, I had visited Espiritu Santo as the guest of a French planter and his family. After only one day of collecting in a local rain forest, I grew ill and was forced to leave on the next flight out.

Now I was part of a team that could stay and discover ants. We made thorough collections along the coasts and into the central ranges of both islands. We found many new species, and were able to place the ant fauna as a whole in the context of that in all of southeast Asia.

We next turned our attention to nearby New Caledonia, and a wholly different challenge. We received word from Hervé Jourdan, an entomologist at the Institute of Research for Development (IRD) in the capital city of Nouméa, that an important species, *Myrmecia apicalis*, thought by some experts to be extinct, had been rediscovered. Would we like to visit the site, search for more of the species, and join in an IRD study of its status?

The New Caledonian bull ant was and remains important for multiple reasons. First, bull ants, named for their large size, aggressive behavior, and powerful sting, are among the definitive insects of Australia. Only one species, *Myrmecia apicalis*, has ever been known outside Australia. Somehow its ancestors—presumably as few as a single pregnant queen—had made it across a considerable span of ocean and evolved its distinctive anatomy in the New Caledonian environment.

The type specimen, on which the Latinized name it was based, was collected in forest on the edge of nineteenth-century Nouméa. The site is now a commercialized suburb. In 1954, I had searched for *Myrmecia apicalis* in vain, taking many trips to other forests around Nouméa and to the north. It seemed likely that the species was extinct.

Not so, Hervé Jourdan wrote to me. An entomologist colleague at IRD had seen the bull ant on L'Ile-des-Pins, a small island 62 miles southeast of Nouméa. Our group now had the opportunity both to make the first study of ants on L'Ile-des-Pins and in the process to learn the status of a very rare ant species.

The condition of any endangered species was an important scientific opportunity all by itself. Almost all previous studies of rarity and extinction have been conducted on vertebrate species: mammals, birds, reptiles, amphibians, and freshwater fishes. Very few had been focused on extremely rare species among the more than one million known insects, spiders, centipedes, snails, and other invertebrates. If we could find and study the remnant *Myrmecia apicalis*, we believed, we might both help to save the ant and add to the meager information available on invertebrate extinction.

On L'Ile-des-Pins, a former penal colony with many of the native habitats intact, we began to search for the bull ants. An early target was a surviving forest of Araucaria, tall conifers of a kind dating back to the early Mesozoic Era. Their height and beauty are responsible for the name of the island. It seemed reasonable to look for ancient ants in an ancient forest, but we discovered instead swarms of the little fire ant *Wasmannia auropunctata*, a notorious invasive species. The species is native to the tropical and subtropical mainland of the New World, and by accidental human activity has been introduced widely through much of the rest of the tropics.

In *The Ants of Florida* (2017), Mark Deyrup described the "insidious nature" of *Wasmannia auropunctata* and the way they dominate the environment of localities they invade. He wrote:

A small number of *auropunctata* workers slowly infil-
trate both the foraging areas and nests of other ants,
going into a defensive posture when challenged. Over
time, however, numbers of *auropunctata* workers can
build up to a level that allows them to band together
and use their sting as well as their repellent and irritat-
ing chemical defense to drive other ants from the foraging arena and kill the inhabitants of the infiltrated
nests. The largest of the conquered colony are gathered
together and probably fed to the *auropunctata* larvae.

In the Araucaria forest, we found that the little fire
ants have displaced almost all other ants and also the
small crustaceans, collembolans, millipedes, and other
invertebrates on which ants of any kind might feed.

Fortunately, Hervé Jourdan knew exactly the locality
on L'Ile-des-Pins where the bull ant had been seen, and we
were all soon gathered at the site. It was radically different
from the Araucaria stands, a beautiful little forest with
dense undergrowth and a low canopy mostly about five
meters high—and fortunately still free of little fire ants.
Jourdan walked in a line close to the edge along the road,
and I walked in parallel on the road, sweeping bushes and
herbaceous vegetation with a net.

It was Jourdan who said, "I think this is the spot." He
stepped into the dense vegetation off the road and soon
shouted, "I've got one." I dashed over to him, oblivious

to the thorny bushes and vines in the undergrowth. And there, walking slowly down the trunk of a small tree, came a *Myrmecia apicalis* worker. I opened a specimen bottle, Jourdan plucked the ant off the tree with thumb and finger and passed it to me. And I dropped it! We looked up and saw a second worker coming down the same tree. Jourdan picked this one off while I swung my sweep net into position. I could tell that the ant was stinging him painfully, but he hung on and when the net was in position, he dropped it into the bottom, and we had our first specimen.

We found its nest at the base of the tree. In the meantime, Christian Rabeling located the nests of two more colonies. Of these none were disturbed, but working together thereafter, night and day, with notebook and camera, we learned a great deal about this rarest of all ant species.

Judging from the size of the nests and the number of foraging workers, the colonies are small, comprising at a guess no more than several hundred. The nests are entered by a single hole leading to a vertical channel to the subterranean chambers we chose not to excavate. The nest entrance is very inconspicuous, made the more so by small amounts of debris placed around or on top of it. The workers forage singly. They leave usually in the morning, climbing up the trunks of the small trees to the canopy to capture or scavenge insects and other invertebrates. They return home, with food, usually toward evening. When disturbed around the nest, they do not attack as do the populous colonies of many of the Australian bull ants. They are relatively timid, cautious when outside the nest, and do not appear to have the numbers or behavior to have an important impact on the ecosystem.

The reverse, however, is true: the ecosystem is killing them. The final, lethal element will be the little fire ants, *Wasmannia auropunctata*, whose dense populations were only a few kilometers away. They are clearly spreading toward the *Myrmecia* locality.

The dark fate of this exquisite little species is entirely up to humanity. *Myrmecia apicalis* can be saved, along with other species still unrecognized, only if the little fire ants are halted and pushed back, and if the woodlands where the New Caledonian bull ant and probably other endangered species yet to be identified live are turned into carefully monitored reserves. L'Ile-des-Pins is today celebrated for the remains of the buildings that once housed a penal colony. It should be given equal or even greater attention for its priceless endangered species.

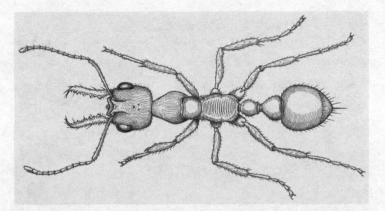

Myrmecia apicalis worker. Drawing of a specimen in the Harvard collection. Collected on L'Ile-des-Pins, New Caledonia. *(Drawing by Kristen Orr.)*

Hervé Jourdan, Christian Rabeling, and Edward Wilson about to board the airplane for travel from Nouméa, New Caledonia, to nearby L'Ile-des-Pins to search for the possibly extinct species of the bull ant *Myrmecia apicalis*, on November 24, 2011. *(Photograph by Jean-Michel Boré.)*

The full expeditionary group, wading through surf in low tide on the way to the site where the very rare *Myrmecia apicalis* had been seen. Left to right: Christian Rabeling, Edward Wilson, Hervé Jourdan, Lloyd Davis, and Kathleen Horton. *(Photograph by Jean-Michel Boré.)*

The first specimen of *Myrmecia apicalis*, caught by Hervé Jourdan, is dropped into a sweep net held by Edward Wilson. The ant at this moment is stinging Jourdan. *(Photograph by Kathleen Horton.)*

Hervé Jourdan and Ed Wilson emerge successful, having located the first nest of the nearly extinct bull ant *Myrmecia apicalis*. *(Photograph by Kathleen Horton.)*

A worker of *Myrmecia apicalis*. *(Photograph by Jean-Michel Boré.)*

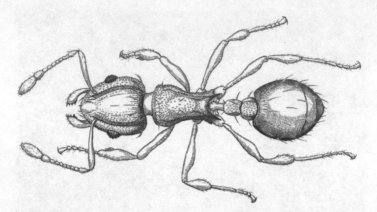

The "killer" species *Wasmannia auropunctata*, an invasive species from tropical South America, responsible for the reduction of ant faunas around the world, including that on L'Ile-des-Pins. *(Drawing by Kristen Orr.)*

The survival of what may be the last population of the endangered bull ant and much of the rest of invertebrate animals is direly threatened by the close approach of swarming "little fire ants," *Wasmannia auropunctata*, seen here in light gray with shiny gasters sprinkled over all of the soil and debris of the forest. *(Photograph by Kathleen Horton.)*

25

LEAFCUTTERS, THE ULTIMATE SUPERORGANISMS

IN 1955, EARLY in my career as a field biologist, I had managed to identify about 175 ant species in a square kilometer of lowland rain forest in Papua New Guinea. This I believed likely to last as a world record. Not so. In later years, twice that number, 355, were collected by Stefan Cover and John Tobin at a single locality in the Amazon.

Forty years after my adventure in New Guinea, I was invited by my friend Thomas Lovejoy to visit a field station he had created near Manaus. I accepted at once. It answered my lifelong dream to study ants in the Amazon—especially in a comfortable, scholarly setting.

The trip to the camp was surprisingly easy. Gone were the fabled days of voyages on sailing ships followed by long treks along machete-cut jungle trails with native bearers and armed guards (and perhaps the ominous sound of drumbeats upriver). In sharp contrast, I found it possible to travel in one day from Boston all the way to Manaus, the capital of Amazon state. I left at dawn, changed flights in Miami, waited in Santo Domingo a short time for refueling, then left on the long continuous haul to Manaus, arriving close to midnight. There followed a short night's sleep before catching an hour-long automobile ride north to the field station.

Anxious to get started, I rose even earlier and walked to a nearby city park, simply out of a desire to find ants and other insects living there. The first ant species I saw was the one I most expected. They were medium-sized and reddish-brown. Many were alone, wandering about. Others ran in files, a few carrying freshly scissored fragments of leaves. They are the ants everyone knows, called leafcutters or fungus growers in English-speaking countries, *saúva* in Brazil, *isaú* in Paraguay, *cushi* in Guyana, *zampopo* in Costa Rica, *wee-wee* in Nicaragua and Belize, *cuatalata* in Mexico, *bibijagua* in Cuba, and town ant or parasol ant locally in Texas and Louisiana at the northernmost part of their range. They are among the ruling insects of the warm temperate and tropical latitudes of the Western Hemisphere.

I had found saúva in urban Manaus. I greeted them quietly, "Hello, little friends."

Taxonomists consider the leafcutter species in Manaus, and elsewhere, to compose the formal tribe Attini, a relatively large group of more than one hundred species spread as a whole throughout the tropics and warm-temperate zones from Argentina to Louisiana. Most attines that cut leaves are placed in the genera *Atta* and *Acromyrmex*.

The *Atta* and *Acromyrmex* leafcutters are notable, above all other animals, for their ability to grow crops of fungi on beds made of chewed fresh vegetation. Extending from this capacity, they are able to build enormously large ant cities. Attine fungus-growers thrive almost everywhere because the vegetation they need is almost endless.

These most amazing of all ants are gardeners, designed in evolution to cultivate symbiotic fungi that are able to flourish only in the care of their attine hosts. Because the ants have almost unlimited space in which to excavate their nests and vast quantities of fresh vegetation on which to grow their fungus crops, they are among the dominant insects throughout their range.

From fresh wild vegetation to rich food crops, how did the attines achieve their breakthrough, and why hasn't it occurred elsewhere in the world by other animal species?

The most likely explanation for the rarity of attine technology is its great complexity. The process by which the attines convert fresh vegetation into an edible fungus is achieved by an intricate cooperation of sophisticated, specialized castes. The castes in turn are physically distinguished by size and allometry (the differential growth of body parts), and the differences they create in the instinctive responses and labor needed to complete each task. The origin of caste systems by allometry is basic to social order throughout the ant world.

Allometry is not only fundamental to ant social behavior. It is also very simple: the larger the ant, the more differential in the relative size of its body parts. (Another expression for allometry is in fact "relative growth.") Most commonly in the ant world, the larger the middle (thorax) and hind (gaster) parts of the body, the larger the proportion of the head.

At the lowest size in a leafcutter colony, the smallest caste, called minors or minims, have body parts similar in proportion to those typical of ants in general. What, then, is the role of the monster-headed supersoldiers at the opposite extreme? They stay deep within the nest, and are usually seen only when the nest is dug open. I thought it possible that these biggest ants are kept as a defense against large predators such as armadillos, bears, and giant anteaters. One day, while staying at a farm on the Magdalena River in Colombia, sprinkled everywhere by leafcutter colonies, I posed a question. Might the supersoldiers be designed to respond to the *smell* of a mammal? How else would they distinguish a deadly foe, as opposed to an ordinary enemy that could be repelled by ordinary soldiers?

To test this hypothesis, I found some of the openings around the edge of nests that funnel fresh air into the deep interior, where the supersoldiers live. Lying on my stomach, I blew into the holes. And within several minutes out stumbled small numbers of the big-headed caste, the first I'd ever seen outside otherwise intact nests.

Supersoldiers have relatively huge heads and sharp heavy mandibles closed by massive adductor muscles that fill most of the voluminous head capsule. With this apparatus, supersoldiers can slice through the chitinous armor of almost any other insect, the skin of mammals—and the leather of your hiking boots.

Minor workers spend most of their time inside the nest, serving as nurses to their immature sisters. They also work as gardeners of the fungus, the principal or sole food of the ants, as it grows on mats of processed vegetation. The large-headed soldiers and especially the still larger-headed supersoldiers have a very different role. They are the heavy-duty military, ready to attack enemies, including anteaters and human farmers who dare to dig into the nest interior.

The minors fill at least one other important role in collusion with their nestmates. The intermediate-sized "medias" construct the nest, process the fragments of newly cut vegetation, and build the fungus mats on which the colony depends. When they are in the field, busy harvesting vegetation for the fungus gardens, they are attacked by small parasitic flies, members of the taxonomic family Phoridae. These agile gnats swoop out of the air and lay eggs on the ants. The eggs hatch into larvae, which burrow through the ants' exoskeleton armor into their bodies, eventually killing them. The phorids emerge as adult flies to start the next life cycle. The ant medias are especially vulnerable when carrying leaf fragments homeward, and the minors serve as their guards. They accompany medias on foraging trips and ride on the leaf fragments, where they act as living fly-whisks. When the phorids fly too close to their targets, the little ants kick at them with their hind legs, driving them away.

The attine leafcutters, like their human occupational counterparts, are able to form dense populations. Their colonies are immense, in fact among the largest known in the whole world of social insects. The mother queen, when inseminated by several males during the nuptial flights, receives 200 to 300 million sperm cells. These she stores in her spermatheca. She pays out sperm cells one by one from the spermatheca during her lifetime of ten to fifteen years. In this time, she gives birth to as many as 150 million to 200 million workers—half the size of the human population of the United States. Partway during the growth to this maximum, they begin to raise virgin queens and males, which in turn are able to disseminate and start new colonies.

The nests built by sexually mature colonies are colossal, perhaps the largest created by an individual or group in the natural world. One typical *Atta sexdens* nest, estimated to be more than six years old, contained 1,920 chambers, of which 238 were occupied by the ants and their fungus gardens. All were interconnected by a complex network of galleries and chambers. The loose soil brought out and piled on the surface during the excavation weighed approximately 40,000 kilograms.

So well-marked and powerful is the division of labor among the members of a leafcutter colony that individual colonies can be reasonably called a superorganism. The expression was first used by the great myrmecologist William Morton Wheeler in 1910, and has been intermittently employed by biologists ever since. In order to survive, a colony, no less than a solitary insect or worker within a colony, must consist of cooperating elements fitted tightly together in the same way that organs are fitted together to function as an organism. The analogy is clear: the soldiers and the minor "fly-whisk" minims are the defense system, the queen is the reproductive organ, other minor workers that tend the gardens are the digestive system, and, finally, the media workers function variously as the brain, hands, feet, and sensory system.

One consequence of this division crafted for unity is that the colony, in addition to each of its members, is a unit of evolution. As a colony changes with time, it competes with other colonies of the same species living around it. The result is natural selection at the colony level. The same process occurs at the level of colony members. Some social traits, such as altruism and bravery in battle, are shaped by "group selection"; in other words, "colony versus colony" leads workers to do what is best for the group. Meanwhile, selection at the individual level leads to selfish behavior, seen when workers, queens, and males compete for space, food, and the right to reproduce.

The ubiquity of the leafcutters, conferred by the prodigious fertility and long life of the queen (thirteen years for one colony kept in the laboratory by Kathleen Horton, at which time males and virgin queens were produced), along with the Stakhanovite labor of the worker daughters, has created major economic problems for their human neighbors. A single colony moving tons of soil cuts agricultural production. One colony can strip a citrus tree of its vegetation overnight, or level an entire family farm garden. And they do so routinely.

The early Portuguese settlers had a phrase for their great hexapod adversaries: *Either Brazil will conquer the ants, or the ants will conquer Brazil.* They were not referring to the furious treetop *Camponotus*, which drop in thousands from their epiphyte-laden gardens to bite and spray formic acid on intruders. People could wave aside the stinging *Pseudomyrmex*, which turn bushes into nettles. Or the *Solenopsis* fire ants, the species whose introduced populations have become scourges in the United States and other countries. Humans can tolerate even the army ants, whose legions in the forests drive almost all animals large and small before them.

A dramatic example of division of labor among the workers of a fungus-growing colony. Media, of intermediate size, carry a newly cut fragment of leaf to use in growing a fungus, which serves as food for the ants. Riding the leaf tips are minors, of smallest size, which guard the larger ants by driving away parasitic flies.

(Photograph by Alex Wild Photography.)

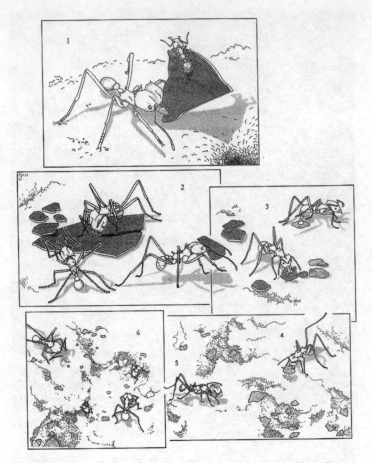

The harvesting of plant fragments and their use as garden substrate to rear fungus for food. *(From B. Hölldobler and E. O. Wilson,* The Leafcutter Ants: Civilization by Instinct, *2011.)*

Clearly the early Portuguese meant another opponent: the leafcutter saúva, destroyers of gardens and crops, spoilers of pastures.

If the saúva could not be conquered, neither could the Brazilians be budged. The result today is a stalemate. It should even be ranked as an environmental blessing. In the magnificent rain forests, savannas, and other terrestrial wildlands, leafcutters are among the premier turners of the soil. In this vital function, and further in their ancient and continuing role as herbivores, they create unique ecosystems and increase biodiversity for habitats as a whole.

The leafcutters are superorganisms that succeed under natural conditions. If not entirely manageable, they are also a source of life and wonder for their human partners.

26

ANTS THAT LIVED
WITH THE DINOSAURS

DURING THE 1950s, I had many conversations with my mentor William L. Brown about the evolutionary ancestry of ants. The thousands of living species we and others studied gave only a few clues to the place and time of the origin of this dominant insect group. Where did ants originate? When? And perhaps even, why?

There were plenty of fossils to study. The richest lode has been the abundant specimens preserved in the Baltic amber, the fossil resin of trees that lived around the present-day North Sea during the Eocene Period, roughly fifty million years ago. Our predecessor at Harvard University, William Morton Wheeler, had assembled and minutely described a large collection. The Baltic amber revealed a great deal of evolution in progress up to the present time, but no clear evidence of the origin of ants in general. The best we could do was extrapolate back in time. The trends in ancestry we found in the Baltic amber fossils.

We guessed what the ultimate ancestors might look like. They would surely be aculeate (stinging) wasps destined to make the giant leap from solitary to social in their mode of life. We needed older fossils—much older than those in the Baltic amber. A scattered few specimens from much older rock deposits suggested that we needed ant-bearing amber specimens from somewhere back in the Cretaceous Period, the uppermost of the Mesozoic Era—the "Age of Reptiles," including dinosaurs.

The break came in 1965, when two amateur collectors, Mr. and Mrs. Edmund Frey of Mountainside, New Jersey, examining fragments of Cretaceous amber from a site in New Jersey, came upon a piece containing two worker ants. They generously donated these scientifically precious fossils to Harvard University. There I had the privilege of studying them for what they were, a clear view of ants from about 90 million years back in time.

What new insight did these ancient couriers bring us? Their external anatomy (at least all that was visible) matched what Bill Brown and I had predicted, but only in part. Some features of the anatomy were more primitive than we had guessed. In other words, the New Jersey Cretaceous ants were a mosaic, combining traits that are probably common to the wasp ancestral to ants and unique to the early traits of ants. The roster was partly as follows:

Mandibles wasplike
Middle part of body antlike
"Waist" antlike to the rear
Antennae intermediate between wasp and ant

I had to invent a two-part Latinized name to formally designate the fossil species. I chose *Sphecomyrma freyi.* The first, the genus name, means "wasp ant" and the second, species name, recognizes and honors the Freys, who found the first specimens.

As other, similar deposits of amber were explored, in Alberta province, Canada, and Myanmar, other sphecomyrmine specimens were found. For a while it seemed as though the missing link in the origin of ants had been located, or we had at least made a close approach. Ants had, it seemed, emerged from a wasp ancestor in the Mesozoic Era along a relatively straight line of evolutionary change, trait by trait.

Then, suddenly—at least as measured in geological time—everything changed. As the collection of Mesozoic fossils increased and the evolutionary progenitors of present-day ants grew more clearly defined, it became apparent that the earliest ants had not evolved in a straight line. Instead, the more successful species experienced an adaptive radiation: a multiplication of species from one or a small number of stocks that became specialized for different niches in the dinosaur-era ecosystems in which they lived.

From one, or a very few, of these radiating species of the late Mesozoic rose the multiple "crown groups" of the great modern ant fauna. As Phillip Barden, Gennady M. Dlussky, David A. Grimaldi, and their fellow analysts have shown, each was the product of an adaptive radiation of its own.

Much of this proliferation—at least, that most readily visible in fossils—was expressed in the structure of the workers' heads. They distinguish the means by which various ant species obtained food as well as defended themselves from enemies. Arguably the most extreme change in the anatomy of the head occurred during the origin of the genus *Haidomyrmex*, first studied by the Russian entomologist Gennady M. Dlussky, who also found the first specimens. From the Greek, the name means "ant from the Land of the Dead," or, as ant biologists say in conversation, "hell ants." Where other ants, including conventional trapjaw species, bring their mandibles together horizontally, like the clapping of two hands, *Haidomyrmex* evolved to turn the apparatus 90 degrees, with the two mandibles fused and moved vertically, up-and-down, so as to close against the labrum, the "upper lip of the head." That this unique device works is suggested by the discovery of one specimen in amber that was captured holding onto a beetle larva.

For amusement, I sometimes ask other naturalists

where they would go and when, if magically they could visit at least for a few hours any place on Earth at any time in the planet's history. My own choice I hold ready: a Mesozoic forest, one hundred million years ago, teeming with ants, including *Haidomyrmex*.

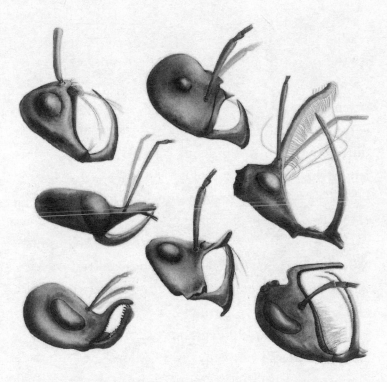

"Hell ants" (haidomyrmecines) of the Mesozoic Era, six species that lived with dinosaurs. Heads shown in side view. The mandibles worked up and down, instead of side to side as in modern ants. *(Drawing by Philip Barden, by permission.)*

ACKNOWLEDGMENTS

I AM GRATEFUL for the contributions of many in the composing of this text, and especially Kathleen Horton of Harvard University and Robert Weil of Liveright Publishing Corporation for their advice and support. My myrmecologist colleagues at Harvard, Stefan Cover and David Lubertazzi, provided excellent corrections and valuable additional information. Kristen Orr provided the precise line-drawing portraits of featured species.

REFERENCES

Barden, P. 2017. Fossil ants (Hymenoptera, Formicidae)—Ancient diversity and the rise of modern lineages. *Myrmecological News* 24: 1–30.

Barden, P., and D. A. Grimaldi. 2016. Adaptive radiation in socially advanced stem-group ants from the Cretaceous. *Current Biology* 26: 515–21.

Brown, Jr., W. L., and E. O. Wilson. 1959. The evolution of the dacetine ants. *Quarterly Review of Biology* 34(4): 278–94.

De Bekker, C., I. Will, B. Das, and R. M. M. Adams. 2018. The ants (Hymenoptera, Formicidae) and their parasites—Effects of parasitic manipulations and host responses on ant behavioral ecology. *Myrmecological News* 28: 1–24.

Delage-Darchen, B. 1972. Une fourmi de Côte-D'Ivoire—*Melissotarsus titubans* Del., n. sp. *Insectes Sociaux* 19(3): 213–36.

Deyrup, M. 2017. *Ants of Florida—Identification and Natural History* (Boca Raton, FL: CRC Press).

Fisher, B. L., and B. Bolton. 2016. *Ants of Africa and Madagascar—A Guide to the Genera* (Oakland, CA: University of California Press).

Frank, E.T., M. Wehrhahn, and K. E. Linsenmair. 2018. Wound treatment and selective help in a termite-hunting ant. *Proceedings of the Royal Society* B 285: 20172457.

Haapaniemi, K., and P. Pamilo. 2015. Social parasitism and transfer of symbiotic bacteria in ants (Hymenoptera, Formicidae). *Myrmecological News* 21: 49–57.

Hölldobler, B., and E. O. Wilson. 1990. *The Ants* (Cambridge, MA: Belknap Press of Harvard University Press).

———. 2011. *The Leafcutter Ants—Civilization by Instinct* (New York: W. W. Norton).

Laciny, A., et al. 2018. *Colobopsis explodens* sp. n., model species for studies on "exploding ants" (Hymenoptera, Formicidae), with biological notes and first illustrations of males of the *Colobopsis cylindrica* group. *ZooKeys* 751: 1–40.

Masuko, K. 1984. Studies on the predatory biology of oriental dacetine ants (Hymenoptera, Formicidae), 1—Some Japanese species of *Strumigenys*, *Pentastruma*, and *Epitritus*, and a Malaysian *Labidogenys*, with special reference to hunting tactics in short-mandibulate forms. *Insectes Sociaux* 31(4): 429–51.

McKeller, R. C., J. R. N. Glasier, and M. S. Engel. 2013. A new trap-jawed ant (Hymenoptera, Formicidae, Haidomyrmecini) from Canadian Late Cretaceous amber. *Canadian Entomologist* 145: 454–65.

Moreau, C. S. 2009. Inferring ant evolution in the age of molecular data (Hymenoptera, Formicidae). *Myrmecological News* 12: 201–10.

Naka, T., and M. Matuyama. 2018. *Aphaenogaster gamagumayaa* sp. nov.—The first troglobiotic ant from Japan (Hymenoptera, Formicidae, Myrmicinae). *Zootaxa* 4450(1): 135–41.

Peeters, C. 2012. Convergent evolution of wingless reproductives across all subfamilies of ants, and sporadic loss of winged queens (Hymenoptera, Formicidae). *Myrmecological News* 16: 75–91.

Peeters, C., and F. Ito. 2015. Wingless and dwarf workers underlie the ecological success of ants (Hymenoptera, Formicidae). *Myrmecological News* 21: 117–30.

Peeters, C., I. Foldi, D. Matile-Ferrero, and B. L. Fisher. 2017. A mutualism without honeydew: What benefits for *Melissotarsus emeryi* ants and armored scale insects (Diaspididae)? *PeerJ* 5: e3599; DOI 10.7717/peerj.e3599.

Schneirla, T. C. 1971. *Army Ants: A Study in Social Organization*, edited by H. R. Topoff (San Francisco: W. H. Freeman).

Wehner, R., and M. V. Srinivasan. 1981. Searching behavior of desert ants, genus *Cataglyphis* (Formicidae, Hymenoptera). *Journal of Comparative Physiology* A 142: 313–38.

Wehner, R., R. D. Harkness, and P. Schmid-Hempel. 1983. *Foraging Strategies in Individually Searching Ants* Cataglyphis bicolor *(Hymenoptera, Formicidae)* (New York: Fischer).

Wheeler, W. M. 1910. *Ants, Their Structure, Development and Behavior* (New York: Columbia University Press).

Wilson, E. O. 1962. The Trinidad cave ant *Erebomyrma* (= *Spelaeomyrmex*) *urichi* (Wheeler), with a comment on cavernicolous ants in general. *Psyche* 69(20): 62–72.

———. 1971. *The Insect Societies* (Cambridge, MA: Belknap Press of Harvard University Press).

———. 2008. One giant leap: How insects achieved altruism and colonial life. *BioScience* 58: 17–25.

Wilson, E. O., and W. H. Bossert. 1963. Chemical communication among animals. In *Recent Progress in Hormone Research, Vol. 19*, edited by G. Pincus (New York: Academic Press), 673–716.

Wilson, E. O., and B. Hölldobler. 1986. Ecology and behavior of the Neotropical cryptobiotic ant *Basiceros manni* (Hymenoptera, Formicidae, Basicerotini). *Insectes Sociaux* 33(1): 70–84.

Zimmerman, E. C. 1970. Adaptive radiation in Hawaii with special reference to insects. *Biotropica* 2(1): 32–38.

INDEX

Page numbers in *italics* refer to illustrations.

ABOUT THE AUTHOR

Edward O. Wilson was born in Birmingham, Alabama, in 1929 and was drawn to the natural environment from a young age. After studying evolutionary biology at the University of Alabama, he has spent his career focused on scientific research and teaching, including forty-one years on the faculty of Harvard University. His thirty-three books and more than four hundred mostly technical articles have won him over one hundred awards in science and letters, including two Pulitzer Prizes, for *On Human Nature* and, with Bert Hölldobler, *The Ants*; the U.S. National Medal of Science; the Crafoord Prize, given by the Royal Swedish Academy of Sciences for fields not covered by the Nobel Prize; Japan's International Prize for Biology; the Presidential Medal and Nonino Prize of Italy; and the Franklin Medal of the American Philosophical Society. For his contributions to conservation biology, he has received the Audubon Medal of the National Audubon Society and the Gold Medal of the Worldwide Fund for Nature. Much of his personal and professional life is chronicled in the memoir *Naturalist*, which won the *Los Angeles Times* Book Award in Science. Wilson has also ventured into fiction, the result being *Anthill*, a *New York Times* best-selling novel. In 2011, PEN America inaugurated the E. O. Wilson Literary Science Writing Award in his honor. Active in field research, writing, and conservation work, Wilson lives with his wife, Irene, in Lexington, Massachusetts.